T0348543

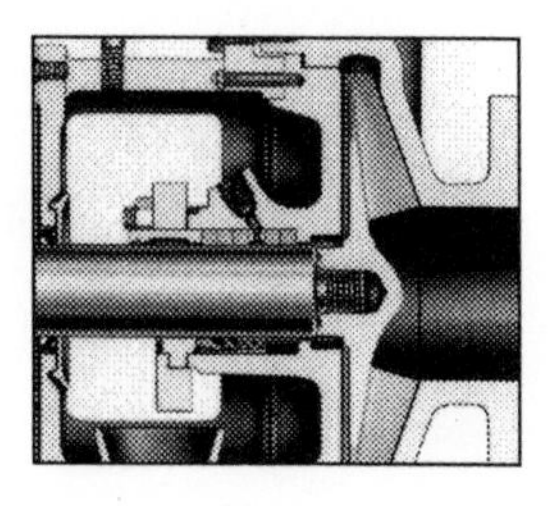

The Practical Pumping Handbook

The Practical Pumping Handbook

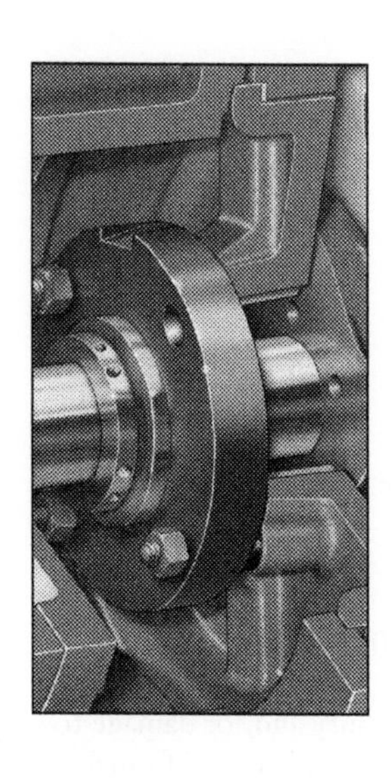

by Ross Mackay

UK Elsevier Ltd, The Boulevard, Langford Lane, Kidlington, Oxford OX5 1GB, UK
USA Elsevier Inc, 360 Park Avenue South, New York, NY 10010-1710, USA
JAPAN Elsevier Japan, Tsunashima Building Annex, 3-20-12 Yushima, Bunkyo-ku, Tokyo 113, Japan

British Library Cataloguing in Publication Data
Mackay, Ross C
The practical pumping handbook
1.Pumping machinery 2.Pumping machinery – Fluid dynamics
I.Title
621.6'9

ISBN 1856174107

Library of Congress Cataloging-in-Publication Data
Mackay, Ross (Ross C.)
The practical pumping handbook / Ross Mackay.
p. cm.
Includes bibliographical references and index.
ISBN 1-85617-410-7
1. Pumping machinery—Handbooks, manuals, etc. I. Title.

TJ900.M25 2004
621.6'9—dc22

2004040454

Published by
Elsevier Advanced Technology,
The Boulevard, Langford Lane, Kidlington, Oxford OX5 1GB, UK
Tel: +44(0) 1865 843000
Fax: +44(0) 1865 843971

Typeset by Land & Unwin (Data Sciences) Ltd., Northamptonshire

Transferred to digital printing in 2007.

Contents

Acknowledgements

A project of this magnitude is never accomplished in a vacuum. The knowledge base from which the information is drawn takes more than one solitary career to acquire. I have been privileged to stand on the shoulders of giants in the pump industry in order to bring you this book.

During my career I have been reminded all too frequently that it is only when you try to teach a concept to others that the depth of your own ignorance becomes apparent. Throughout the years I have been rescued from those depths by more people than either my memory allows me to name, or sufficient space is available to identify here. They are the many thousands of associates, clients and students with whom I have had the privilege of working, and who have challenged me and kept me growing along the years.

I am especially grateful to those generous and brave souls who critiqued various parts of this book and, in doing so, made it more accurate and more complete. They are, Ed Avance of Advanced Sealing, Darren Bittick of International Paper, Brian Dahmer of MRC Bearing Services, Kevin Delaney of Kevin Delaney M.S., Dave Djuric of Alberta Pacific, Neil Flanagan and Dave Meloche of ProSpec Technologies, Jerry Hallam of BP Chemical, Dave Meister of Gorman-Rupp, Mike O'Neill of Unilever and Mark Weare of Weyerhaeuser.

My thanks also goes out to The Hydraulic Institute and The American Petroleum Institute for allowing us to incorporate a small portion of the extensive reference material they work so hard at making available through their own, more exhaustive publications.

For my ability to write what I mean and mean what I write (O.K., most of the time) my thanks go out to my high school English teacher, Jimmy Ross and, more recently, to Jane Alexander who has applied the polish to most of my articles in recent years, and thus, much of this book.

To my wife, Margaret, and our children, Lesley, Laura, John Paul and Paul, thank you for your love and support throughout a checkered career.

So much help, so freely given by so many, made this an enjoyable project. Undoubtedly I made more mistakes than my colleagues were able, through their vigilance, to eliminate. The remaining errors are my responsibility entirely.

Dedication

To my grandsons, Matthew and Michael,

who bring joy to my life every single day.

May you always do what you love and love what you do.

About the author

Ross Mackay specializes in helping companies reduce their pump operating and maintenance costs. His unique breadth of experience in pumps, seals and pumping systems has been gained through extensive international exposure to industry in over 30 countries around the world.

Through his renowned Mackay Pump School he has trained thousands of Operations and Maintenance Engineers and Technicians in the Science of Pumping Reliability. These clients come from a wide range of industries that are dependent on the efficient movement of liquids, such as Pulp and Paper, Power, Petro-Chemical, Water and Waste Treatment, and many others.

The Mackay Pump School is a comprehensive Reliability Training Program focused on improving trouble-shooting skills to increase pump reliability and thus eliminate ongoing and repetitive pump failure. Implementation of the ideas gained from this school has saved end users millions of dollars in increased efficiency and reliability. Mr. Mackay accepts a number of engagements every year to conduct in-house training programs on pump reliability and troubleshooting.

Ross Mackay is also the author of the video learning program, "A Practical Approach to Pumping" that explores the three vital areas of pump mechanics, system hydraulics and seal operation, integrating them to simplify root cause analysis and effective trouble-shooting. Highly recommended for those seeking a further appreciation of process pump design and operation. He also writes a monthly email newsletter, 'The Pumpline' that provides brief tips and techniques on pumping reliability.

A graduate in Mechanical Engineering from Stow College of Engineering in Glasgow, Scotland, and a member of the PAPTAC,

TAPPI and the AWWA, he has been associated with such companies as Weir, BW/IP, Bingham and Chesterton.

As a respected authority on pumps, with dozens of feature articles in major industry magazines, he has an enviable international reputation and is a popular speaker at major conferences. He is also in great demand as a keynote and after-dinner speaker who brings his enthusiasm and love of laughter to every audience while leaving them with life skills that improve their performance and productivity.

Ross Mackay Associates Ltd.
4 Simmons Crescent
Aurora, Ontario, Canada, L4G 6B4
Telephone: 1-905-726-9587
Email: info@practicalpumping.com
Web Site: www.practicalpumping.com

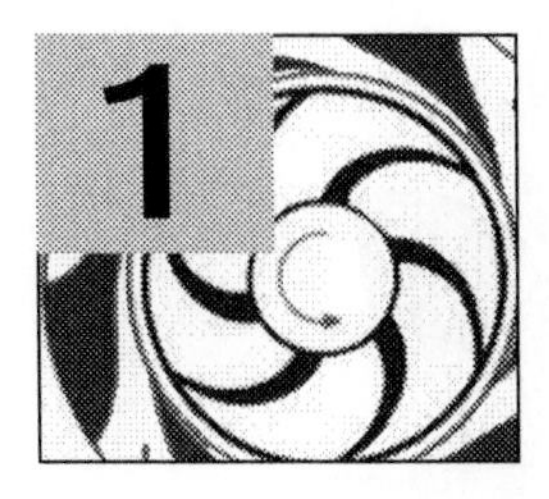

1 Centrifugal pumps

1.1 The pump

A pump is an item of mechanical equipment that moves liquid from one area to another by increasing the pressure of the liquid to the amount needed to overcome the combined effects of friction, gravity and system operating pressures. In spite of the wide divergence of pump types available, over 80% of all pumps used in industry are of the single stage, end suction, centrifugal pump.

The centrifugal pump moves liquid by rotating one or more impellers inside a volute casing. The liquid is introduced through the casing inlet to the eye of the impeller where it is picked up by the impeller vanes. The rotation of the impeller at high speeds creates the centrifugal force that throws the liquid along the vanes, causing it to be discharged from its outside diameter at a higher velocity. This velocity energy is converted to pressure energy by the volute casing prior to discharging the liquid to the system.

Two pump types are more commonly used than all the others put together. They are the ANSI pump that is designed and built to the standards of the American National Standards Institute, and the API pump that meets the requirements of the American Petroleum Institute Standard 610 for General Refinery Service. While other countries have their own designations, such as the International ISO Standards, the German DIN Standards and the British BS Specifications, the pump styles are still very similar to either the ANSI or the API pump.

Over the years, ANSI designs have become the preferred style of end suction pumps, not only for chemical process applications, but also for water and other less aggressive services. The ANSI Standard provides for dimensional interchangeability of pumps from one manufacturer to another.

Figure 1.1: ANSI process pump (Reproduced by permission of Flowserve Corporation)

The API pump is almost the exclusive choice for applications in the oil refining and associated industries, where it handles higher temperatures and pressure applications of a more aggressive nature. While API specifications also deal with some vertical shaft models, the horizontal style is the more widely used design.

These single stage pumps are both designed with a radially split casing to accommodate a pullout arrangement at the back for ease of maintenance. The major difference between the two styles is reflected in the casing pressure design ratings, which are as follows:

ANSI Pump Rating = 300 PSIG at 300° F

API Pump Rating = 750 PSIG at 500° F

In view of these figures, it is apparent that the API pumps should be considered for higher pressure and temperature services than the lighter duty ANSI pump.

1.2 Applications

In considering the different types of liquids handled by these pumps, the various applications are frequently classified in the following categories:

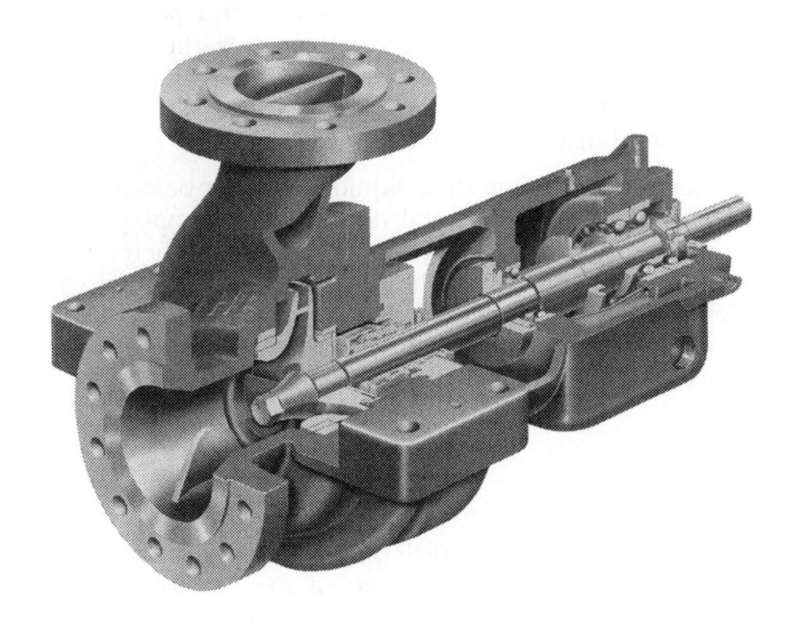

Figure 1.2: API process pump (Reproduced by permission of Flowserve Corporation)

- Hydrocarbons,
- Chemicals,
- Slurries, and
- Water.

Hydrocarbons are petroleum-based products that are usually further classified as light, intermediate or heavy. At atmospheric pressure and temperature, light hydrocarbons tend to vaporize, intermediate hydrocarbons are liquid, and heavy hydrocarbons are highly viscous or even solid.

Chemicals include strong acids, alkalines or oxidizing agents that are destructive to both equipment and the environment. They can also be dangerous to plant personnel if allowed to leak.

Slurries constitute a mixture of solid particles in a liquid that is usually water. They come in a wide variety of products and waste material, and the pumps required in these services will be discussed in Chapter 8.1.

Water and water type liquids (including some mild chemicals) are generally easy to handle, and are not detrimental to either equipment or the environment.

Many of these more aggressive liquids can produce toxic fluid exposure and vapors if they are allowed to leak out of a pump. For example, vapor release is a common danger with hydrocarbons that vaporize at atmospheric conditions or other chemicals that may be exposed to very high operating temperatures. If a vapor release is exposed to a spark, the vapor cloud may even explode or catch fire.

Consequently, in handling these liquids, we must be extremely aware of much more than environmental damage and pumping efficiency. We must also be very conscious about personal safety. Therefore, the choice between the ANSI pump and the API pump must take into account the specific fluid properties, as well as the operating conditions. The main difference between these pumps is predominantly a result of the differences in casing design.

1.3 Pump cases

Both pump styles have a radial split casing, and most smaller pump cases employ a single volute design of the interior passages. This is particularly evident with low-flow rates and lower specific speeds of the impeller.

As shown in Figure 1.3, the impeller is offset within the volute design and that point in the casing that is closest to the impeller is referred to as the 'cut-water'. In a counterclockwise direction from this point, the scroll design of the casing wall steadily moves away from the impeller around its perimeter. This develops the pump capacity throughout the rotation until it exits the discharge nozzle located on the pump centerline.

Figure 1.3: Single Volute Casing

As the wall of the casing retreats from the impeller, the area of the volute increases at a rate that is proportional to the rate of discharge from the impeller, thus producing a constant velocity at the periphery of the impeller. This velocity energy is then changed into pressure energy by the time the fluid enters the discharge nozzle.

The peculiar shape of the volute also produces an uneven pressure distribution around the impeller, which in turn results in an imbalance of the thrust loads around the impeller and at right angles to the shaft.

This load must be accommodated by the shaft and bearings, and much has been discussed on this problem in recent years.

The resultant unbalanced load is at its maximum when the pump is run at the shutoff condition. It gradually decreases as the flow rate approaches the Best Efficiency Point (BEP). If the pump operates beyond the BEP, the load increases again, but in the opposite direction on the same plane. Examination of the resultant shaft deflection problems has indicated that the radial plane on which the out-of-balance load acts is approximately 60° counterclockwise from the cut-water of the volute.

Most of the larger API pumps are produced with a double volute design to reduce these loads on high-flow and high-head units. (See Figure 1.4.) This is accomplished by balancing the opposing out-of-balance loads from each volute. While the cost of this is a slight reduction in efficiency, it is considered a small price to pay for the increased reliability that ensues.

Figure 1.4: Double volute casing

Another casing feature found in many API pumps is the top suction/top discharge arrangement, where the suction nozzle is located at the top of the casing adjacent to the discharge nozzle, rather than on the end.

On the vertical inline design, the suction nozzle is once again on the side, but now it is opposite to the discharge nozzle, thus creating the 'inline' appearance. The drawback of this design is that, for many of these pumps, the Net Positive Suction Head (NPSH) required is often considerably greater than it would be in the end suction arrangement. More NPSH is needed in order to accommodate the friction losses in the tortuous path from the suction flange to the eye of the impeller.

These vertical inline pumps do provide the considerable advantage of eliminating the baseplate/foundation requirements and costs, as well as minimizing the footprint area required for their installation. The older designs of inline pumps, many of which are still in service throughout the world, do not include a bearing for the pump shaft and relied solely on the motor bearings. Newer designs as shown in Figure 1.5 now provide the additional stability and reliability of a pump bearing located between the stuffing box and the coupling.

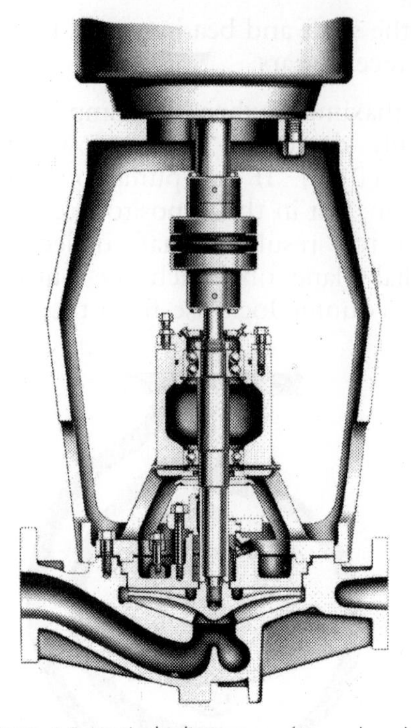

Figure 1.5: Vertical inline pump (Reproduced by permission of Goulds Pumps, ITT Industries)

Figure 1.6: Diffuser casing

1.3.1 Diffuser casings

Another design style incorporates a circular casing with a diffuser which has the interior passages needed to transfer the velocity energy to the pressure energy prior to discharge from the casing. In this design, the impeller runs concentrically within the diffuser and the casing. This arrangement is used extensively in multistage pump designs in both vertical and horizontal configurations.

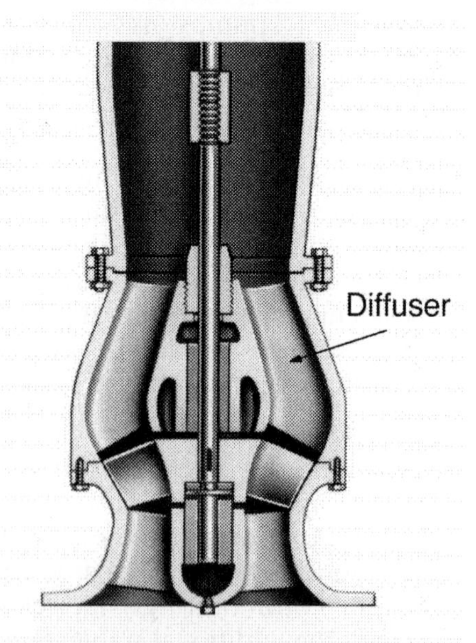

Figure 1.7: Vertical turbine bowl assembly (Reproduced by permission of Goulds Pumps, ITT Industries)

Diffuser vanes are used in a slightly different arrangement, yet with the same purpose, in a vertical turbine pump. Where the impeller discharges into the bowl assembly casing, the diffuser vanes in that casing guide the liquid into the eye of the next stage or into

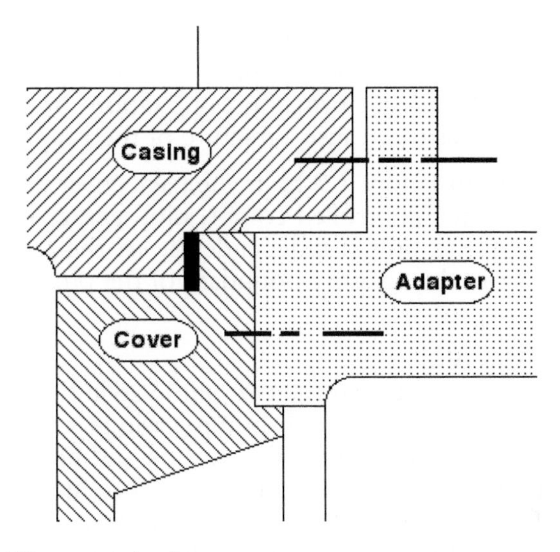

Figure 1.8: Typical ANSI pump casing/cover

the discharge column. Further details on vertical pumps will be found in Section 9.2.

1.3.2 Back cover arrangements

One of the major differences between the ANSI and API pump casings is in the manner in which the back cover is secured to the casing.

In the ANSI design shown in Figure 1.8, the back cover and gasket are held against the pump casing by the bearing frame adaptor, which is most frequently supplied in cast iron. This usually results in a gap

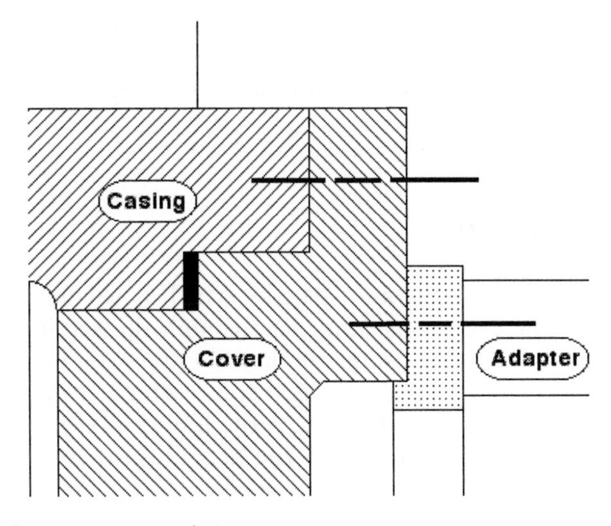

Figure 1.9: Typical API pump casing/cover

between the mating faces of the frame adaptor and the pump casing that has the potential to permit uneven torquing of the bolts. In the event of a higher-than-normal pressurization of the casing by the process system, this may cause a fracture of the adaptor.

The API design in Figure 1.9 bolts the back cover directly to the casing and uses a confined controlled compression gasket with metal to metal fits. The adaptor is bolted independently to the back cover and does not play a part in the pressure boundary of the pump casing.

1.3.3 Mounting feet

Another difference between the two pump styles is the configuration of the mounting feet. All ANSI pump casings are mounted on feet projecting from the underside of the casing and bolted to the baseplate. If these pumps are used on high-temperature applications, the casing will expand upwards from the mounting feet and cause severe thermal stresses in the casing that will detrimentally affect the reliability of the pump. Operation at lower temperatures will not be affected by this feature.

On the other hand, API pumps are mounted at the horizontal centerline of the casing on feet projecting from each side of the casing and bolted to pedestals that form part of the baseplate. This arrangement provides the API pump with the advantage of being able to operate with pumpage at elevated temperatures. As the pump comes up to temperature in such cases, any expansion of the metal will be above and below the casing centerline, and will exert minimal amounts of stress to the casing, thus contributing to optimum reliability of the pump.

The ability to handle higher temperature services is also evident in the bearing housings of the API pumps, which tend to be much more robust in design and also accommodate cooling jackets with a greater capacity of cooling water.

1.4 The impeller

The impeller is secured on a shaft by which it is rotated. The liquid is delivered to the eye of that impeller through the suction nozzle located at the end of the pump. After the liquid enters the eye of the impeller, the rotation creates the centrifugal force which moves the liquid out along the vanes to the perimeter. As the liquid moves towards the outer diameter, its velocity increases to the maximum that it achieves when it leaves the outer diameter of the impeller.

In order to produce different relationships between the flow rate and pressure, impellers are designed in various configurations. This allows them to be selected to meet the governing criteria of the specific service in which they will operate.

1.4.1 Specific speed

The form and proportions of impellers vary with a non-dimensional design index that is referred to as Specific Speed. This is defined as the revolutions per minute at which a geometrically similar impeller would run if it were of such a size as to discharge one gallon per minute against one foot of head. It has been found that the ratios of the major dimensions of the impeller vary uniformly with specific speed. Consequently it is useful to the pump designer in predicting the proportions required when comparing impellers.

$$Ns = \frac{N\sqrt{Q}}{H^{0.75}}$$

When calculating the specific speed, it must be noted that all values must be taken at the best efficiency point (BEP), at the maximum impeller diameter and at the rated speed of the pump.

If an impeller has a low value of Specific Speed (say 500 to 1,000), it is identified as a Radial Vane Design in which the liquid makes a 90° turn from the horizontal entrance flow to radially from the shaft. These impellers generally deliver a relatively low flow rate against a high head, but with a low efficiency. They are used extensively in such services as boiler feed pumps where the main function is to pressurize the water before it goes into the boiler.

The high values of Specific Speed (5,000 and up) identify Mixed Flow and Axial Flow Impellers where the flow path varies minimally from being parallel to the axis of the shaft. These impellers generally deliver a

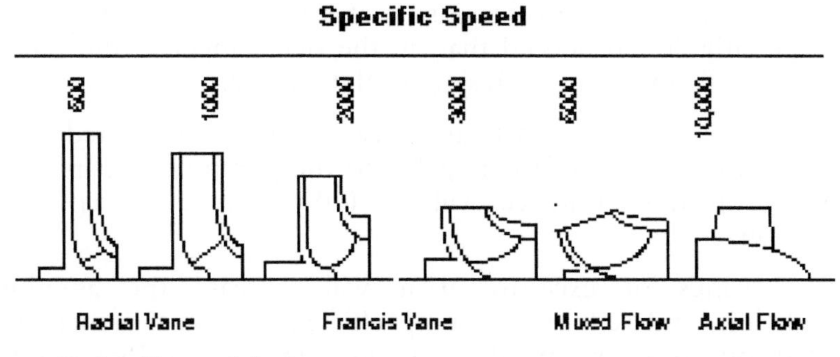

Figure 1.10: Specific speed chart

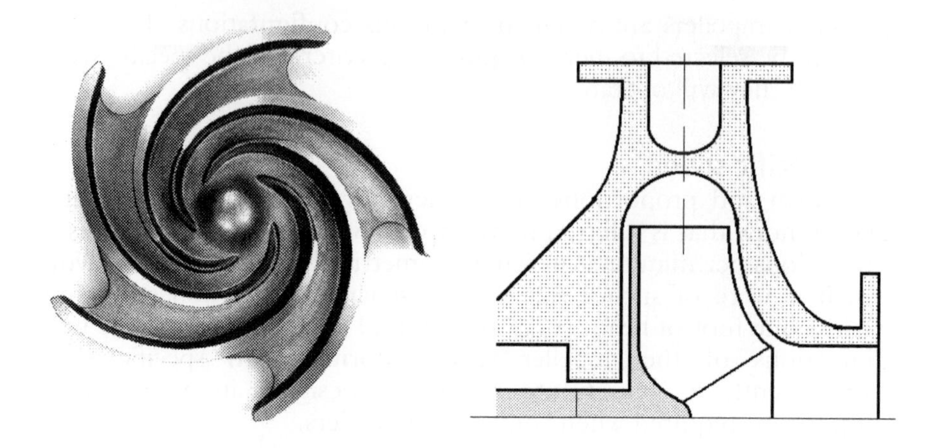

Figure 1.11: Open impeller (Reproduced by permission of Goulds Pumps, ITT Industries)

Figure 1.12: Recirculation with an open impeller

relatively high flow rate against a low head, and with a high efficiency. These impellers are frequently found in large vertical propeller pumps where the main function is to move as high a volume of water as possible at relatively low pressures. Irrigation services and main intake pumps for power stations and paper mills use these impellers extensively.

The most common impeller designs have Specific Speed values that fall in the range of 1500 to 3000, and are commonly referred to as Francis-Vane impellers. This group generally delivers a fairly wide range of Flows at medium Heads and consequently is used extensively in water and general service process pumps.

1.4.2 Open and closed impellers

The most distinctive difference between impellers in a process pump is whether or not they are 'open' or 'closed'. Both types are widely used.

The open impeller has no shrouds on at least one side of the impeller. Consequently, it can be said that, in the open impeller, the vanes are easily visible from one or both sides of the impeller.

Recirculation with an open impeller design is restricted by the proximity of the front of the impeller to the pump casing. In a typical ANSI pump this clearance will be 0.015 ins. on a cold liquid application.

Wider settings or excessive wear will increase the amount of recirculation, and reduce the pump efficiency. This clearance can be adjusted on most pumps through an arrangement located at the

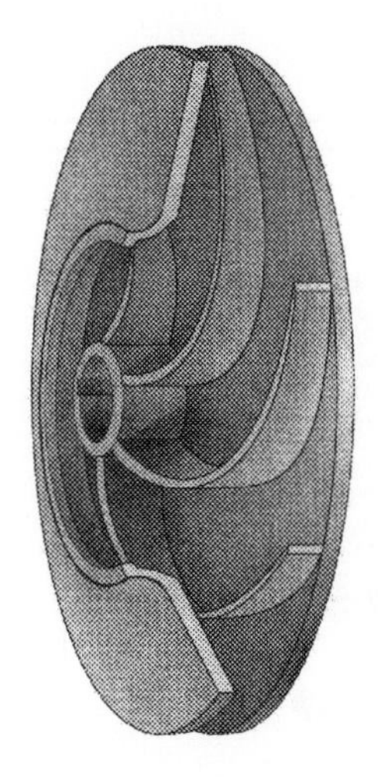

Figure 1.13: Closed impeller

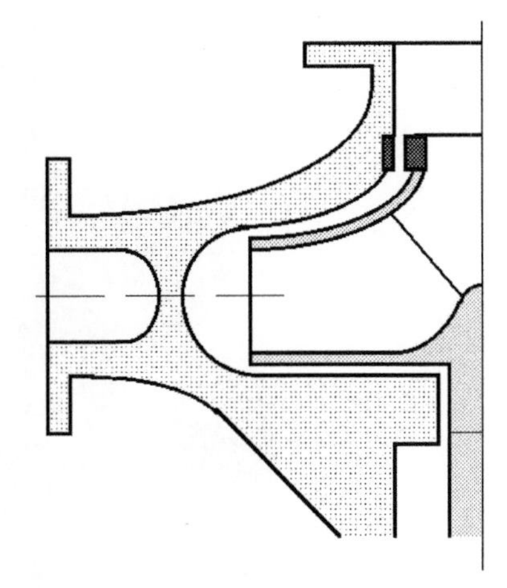

Figure 1.14: Recirculation with a closed impeller

coupling end of the bearing housing. However, this adjustment method should not be used when a mechanical seal is installed and locked in place as it will damage that seal.

Pump-out vanes on the back of the open impeller will assist in balancing the axial thrust and reducing the pressure in the stuffing box.

In a closed impeller, the liquid passages are contained within the impeller by shrouding the impeller vanes. This arrangement is generally considered to be more efficient than the open impeller design as it tightly contains the flow of liquid from the eye of the impeller, all the way through to the periphery. However, the hydraulic efficiency of a pump in service is primarily affected by the amount of recirculation that takes place from the high pressure perimeter of the impeller to the low pressure eye area. As wear takes place in the critical areas and opens the critical clearances, recirculation is increased and the efficiency of the pump will decrease, thus raising the power draw.

Closed impellers will often use wear rings to limit the clearance and to reduce the recirculation. When one or both rings wear, the clearance will open up, the recirculation will increase and the efficiency will drop. When the efficiency reaches an unacceptable level, the rings should be replaced in the 'as new' condition.

Wear rings are also used on the back of the impeller to assist in axial hydraulic balance of the rotating element. Balance holes in the impeller can assist by equalizing the pressures behind the impeller and at the eye area. This arrangement will also contribute to reducing the pressure in the stuffing box.

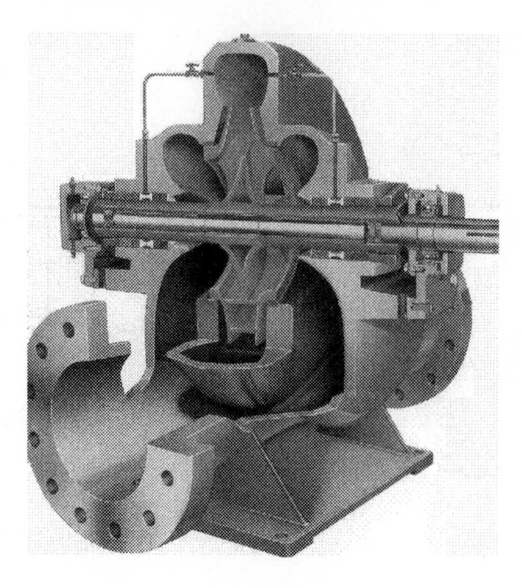

Figure 1.15: Double suction pump (Reproduced by permission of Fairbanks Morse – Pentair Pump Group)

1.5 Double suction pumps

In applications where large volume flows are required, a single stage horizontal double suction pump may be used. They are also the preferred design when a high degree of reliability is required such as in remote pipeline applications. These pumps consist of a single double suction impeller where the liquid enters the impeller from both sides simultaneously and thus creates a high degree of axial balance in the rotating element. It also contributes to the fact that both stuffing boxes only see the suction pressure of the pump.

The pump casing is horizontally split along the axis of the shaft. This permits removal of the rotating assembly without disturbing suction and discharge piping or the driver mounting. The lower half of the casing includes the heavy mounting structure.

1.6 Materials of construction

Most pump manufacturers can provide their pumps in a wide assortment of materials, the selection of which depends on the operating stress and effects, as well as the type of wear from the product being pumped. The most common materials used in end suction centrifugal pumps are as follows:

- Cast and Ductile iron
- Bronze
- Carbon and low alloy steels such as 4140.
- Chrome steels such as 11%, 12% or 13%
- Martenistic stainless steels in the 400 series.
- Precipitation hardening stainless steels like 17-4 PH.
- Austenitic stainless steels like the 300 series or alloy 20.
- Duplex stainless steels such as CD4MCu
- Other more exotic alloys such as Hastelloy, Titanium, etc.

Iron and bronze pumps are widely used in general service applications such as may be found in the water and waste treatment facilities, steel mills and marine applications. The stainless steels are used for a range of corrosive solutions and are suitable for many mineral acids at moderate temperature and concentrations. Further information on pump materials of construction can be found in Chapter 15.

1.6.1 Nonmetallic pumps

Nonmetallic pumps also play a major role in the movement of chemicals, and a number of plastics are used as pump linings as well as complete pump units because they offer the corrosion resistance of the more expensive metals at a fraction of the cost. However, they do have strength limitations that may inhibit their use in certain areas. An extremely high degree of chemical resistance can be found in the fluorocarbon resins such as polytetrafluoroethylene (PTFE). Where additional strength and chemical resistance is needed, a variety of fiber-reinforced plastics (FRP) are available.

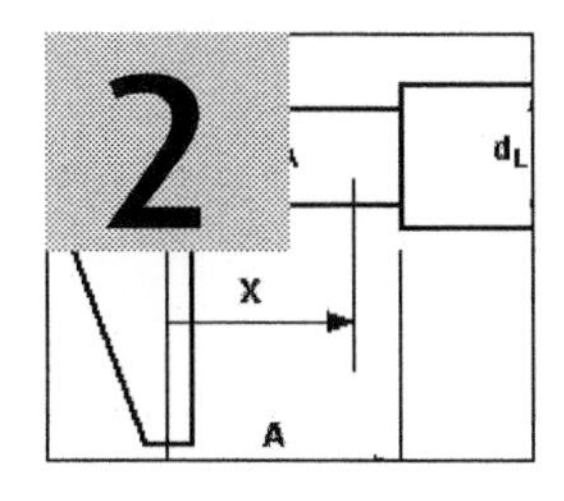

Pump hydraulics

2.1 The pressure-head relationship

In considering the amount of pressure energy required from a pump, all the various aspects of energy in the system on both sides of the pump must be considered. As these energy levels are customarily identified in 'pressure' terms (such as pounds per square inch) or in 'head' terms (such as feet of head) it is important to be comfortable with the relationship between these two important terms.

All pressures can be visualized as being caused by a column of liquid which (due to its weight) would produce a pressure at the bottom of that column.

To exert a pressure of one pound per square inch at the base of a column of water at 60° Fahrenheit, with a specific gravity of 1.0, that column must be 2.31 feet high. To exert a pressure of 14.7 pounds per square inch at its base, that column must, therefore, be 34 feet high. This assumes that there is no external pressure being exerted on the top of that column of water.

Therefore, it can be assumed that if a tank of water at the same temperature is open to atmosphere at sea level, it will have a pressure on its surface of 14.7 p.s.i. or, in other terminology, 34 feet of head.

Therefore, in that same tank of water at the same temperature, the pressure existing at any point in the liquid will be the sum of the weight of the liquid above that particular point, plus the pressure on the free surface of the water.

In other words, the total head being exerted at the bottom of a storage tank of water, 15 feet deep, and open to atmospheric pressure at sea level, will be 15 feet depth plus 34 feet of atmospheric pressure. This equals a total head of 49 feet.

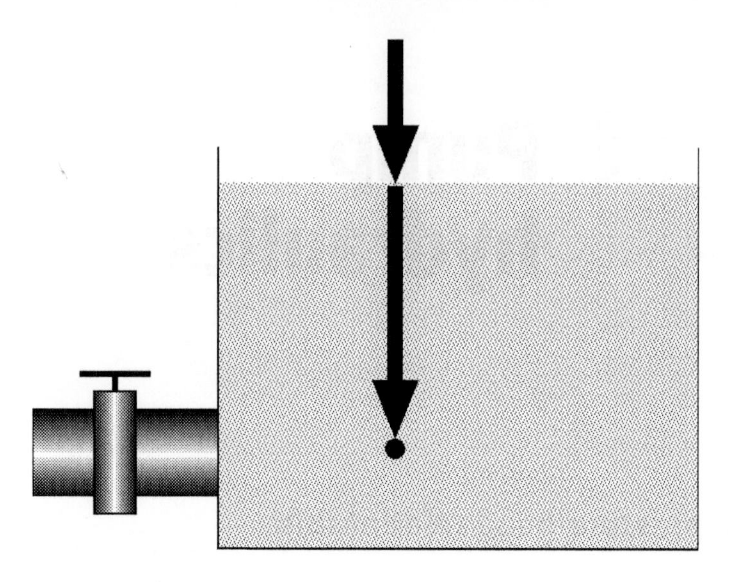

Figure 2.1: Total head in tank

In more general terms, the relationship between Pressure and Head when dealing with water at 60° F, is as shown, when Pressure is in PSI and Head is in Feet.

$$\text{Head (in feet)} = \text{Pressure (in p.s.i.)} \times 2.31$$

When other liquids are in use, however, it is necessary to consider the different densities of these liquids. The ratio of the Density of any liquid to the Density of Water at 60° Fahrenheit is called the Specific Gravity. Consequently, the following formula will apply.

$$\text{Pressure (in p.s.i.)} = \frac{\text{Head (in feet)} \times \text{Specific Gravity}}{2.31}$$

2.1.1 Pressure terminology

Of the various terms used to identify pressure in the pumping field, all of these are compared to some specific base pressure, and are clarified as follows:

Absolute Pressure relates to the complete absence of pressure, as in a perfect vacuum. It is the amount by which the stated pressure exceeds a perfect vacuum.

Gauge Pressure is related to atmospheric pressure and is the amount by which the stated pressure exceeds atmospheric pressure.

e.g. Absolute Pressure = Atmospheric Pressure + Gauge Pressure

Vacuum is also relative to atmospheric pressure, but it is that amount by which the stated pressure is less than atmospheric pressure.

e.g. Absolute Pressure = Atmospheric Pressure – Vacuum Gauge Reading

Differential Pressure is the difference in pressure between two points in a system. More specifically, it is used to identify the difference between the suction and discharge pressures of a pump when referred to the same datum and expressed in feet of head.

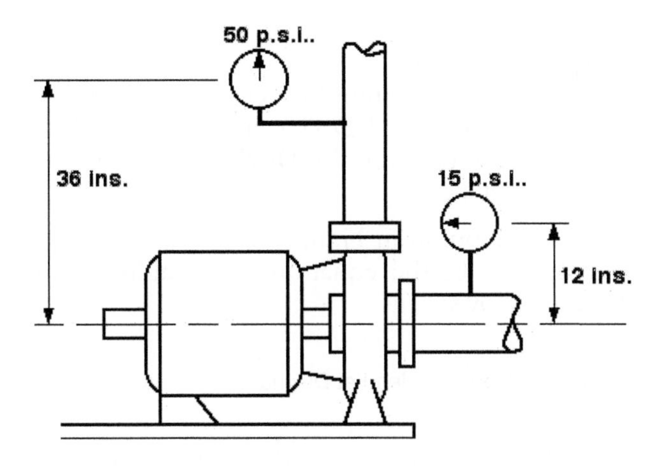

Figure 2.2: Differential pressure measurement

2.1.2 Pumping pressure

To relate this information more closely to the pump operation, consider three identical pumps with the same impeller diameter, and running at the same speed. Each pump is handling a liquid of different density to the others.

- the Kerosene has a Specific Gravity of 0.8%
- the Water has a Specific Gravity of 1.0
- while the Sulphuric Acid has a Specific Gravity of 1.8

This means that the Kerosene is 20% lighter than Water, while Sulphuric Acid is about 80% heavier than Water. In spite of this, all 3 pumps will develop the same Head of 100 feet when running at the same speed. The Discharge Pressure of the pumps is quite different however, because of the different densities of the various liquids.

Liquid	S.G.	Head	Pressure
Kerosene	0.8	100ft.	34.63 p.s.i
Water	1.0	100ft.	43.29 p.s.i.
Sulphuric Acid	1.8	100ft.	77.92 p.s.i.

Figure 2.3: Pressure-head table

This is why it is much simpler to discuss the performance of a pump in terms of Head rather than Pressure. The use of Head makes the pump curve applicable to every liquid regardless of Density.

2.1.3 Total dynamic head

The energy added to the system by a centrifugal pump is referred to as the Total Dynamic Head (T.D.H.) and can be calculated from the difference in pressure between the Discharge side of the pump and the pressure on the inlet side.

For example, if the pump in Figure 2.2 is pumping cold water with a specific gravity of 1.0, the equation for establishing the Differential Head will be as follows:

$$\text{Total Dynamic Head} = \text{Head at Discharge} - \text{Head at Suction}$$

$$\text{T.D.H.} = \left[\left(\frac{50 \times 2.31}{1.0}\right) + \left(\frac{36}{12}\right)\right] - \left[\left(\frac{15 \times 2.31}{1.0}\right) + \left(\frac{12}{12}\right)\right]$$

$$= 118.5 - 35.65$$

$$= 82.85 \text{ feet}$$

2.2 Performance curve

The Total Energy Output of a pump is a combination of the Total Dynamic Head and the Flow Rate, and the relationship between the two is shown on a pump performance curve.

For a simplified explanation of how a pump curve is developed, consider a Centrifugal Pump discharging into a straight vertical pipe. Eventually the liquid will reach a maximum level, beyond which it is unable to move. This can be considered as the maximum Head the pump can develop and, at this point, the pump will continue to run, but will be unable to push the liquid any higher in the pipe. Under these conditions, liquid is agitated in the pump casing, but there is no flow passing through the pump, therefore the flow rate is Zero at this Maximum Head.

If we cut holes in the discharge pipe at progressively lower levels, the Head is effectively reduced, and the pump will develop an increasing flow rate. By graphically depicting these results as shown in Figure 2.4, the characteristic pump performance curve is drawn.

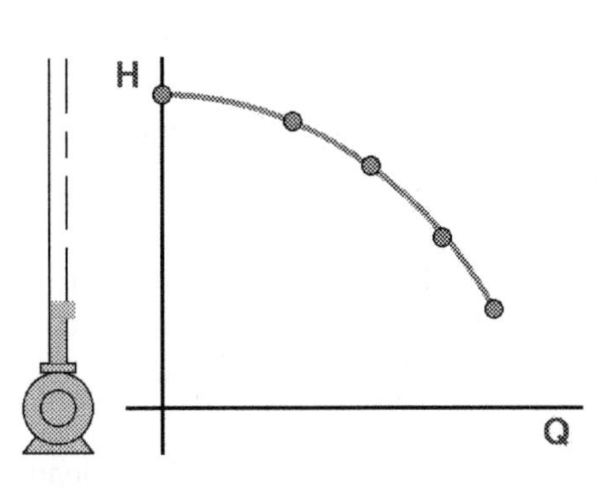

Figure 2.4: Effect of reducing head on capacity

It should be noted that this curve is not completed down to Zero Head, as a centrifugal pump does not operate reliably beyond a certain Capacity. Consequently, at that point, the curve is usually discontinued.

This curve identifies the Capacity which this pump can develop, and the Total Head it can add to a system and is, therefore, usually referred to as the 'Head–Capacity' curve. In addition, when depicted as in Figure 2.5, it is frequently referred to as the 'Single Line Curve' as it displays the performance of the pump when one particular impeller diameter is installed and the pump is run at a predetermined speed.

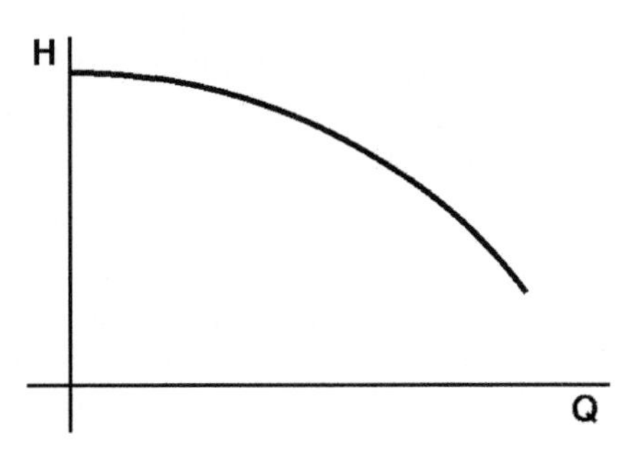

Figure 2.5: Single line pump performance curve

2.2.1 Efficiency

The Head-Capacity capability requires a certain amount of power to be supplied to the pump. The actual quantity of power required will be dependent on how efficiently the pump operates.

The Efficiency represents the percentage of the total power used in the direct development of the Capacity and the Total Head. In general terms, Efficiency is the work produced by a machine divided by the work supplied to that machine.

$$\text{Efficiency} = \frac{\text{Work out}}{\text{Work in}}$$

For centrifugal pumps, it is the Capacity multiplied by the Total Head and divided by the Power Input. When working in Gallons per Minute and Feet of head, the formula is as follows:

$$\text{Efficiency} = \frac{\text{USGPM} \times \text{Head (in feet)} \times \text{Specific Gravity}}{\text{Horsepower} \times 3960}$$

When selecting a pump, we usually know the Efficiency and need to find the horsepower in order to size the driving motor; therefore, the equation is used as follows:

$$\text{H.P.} = \frac{\text{USGPM} \times \text{Head (in feet)} \times \text{Specific Gravity}}{\text{Efficiency} \times 3960}$$

In this equation an efficiency of 67% would be identified as 0.67.

With the same Head-Capacity curve for the maximum impeller diameter, we can establish another vertical axis and draw in the pump Efficiency curve.

The flow rate, at which the highest point on the efficiency curve is achieved, is known as the Best Efficiency Point. This BEP is the most stable operating condition for that pump.

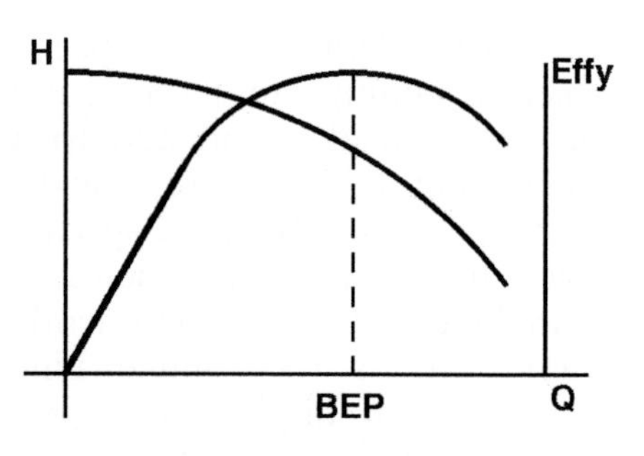

Figure 2.6: Efficiency curve

2.2.2 Net positive suction head

To ensure that Head, Capacity and Efficiency are fully developed by the pump, a suitable hydraulic condition is required at the inlet to the pump. This condition is referred to as the Net Positive Suction Head Required (NPSHR) and can be drawn against another vertical axis. The NPSH required by the pump (NPSHR) must be made available from the system (NPSHA) in order for the pump to fully develop the Head-Capacity at the efficiency shown on the curves.

The performance curve shown on Figure 2.7 represents the total hydraulic capability of the pump, when operating at one particular speed. It shows

- The amount of capacity the pump can develop.
- The level of total head the pump can add to the system.
- The efficiency with which this is accomplished.
- The minimum amount of NPSH required ensuring the pump can develop that total hydraulic capability.

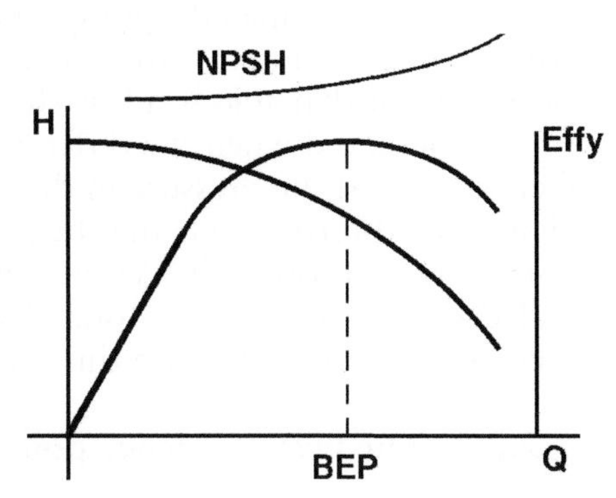

Figure 2.7: NPSH curve

2.2.3 The composite pump performance curve

A Composite Curve displays the total capability of the pump with various impeller diameters from the largest to the smallest as shown on Figure 2.8.

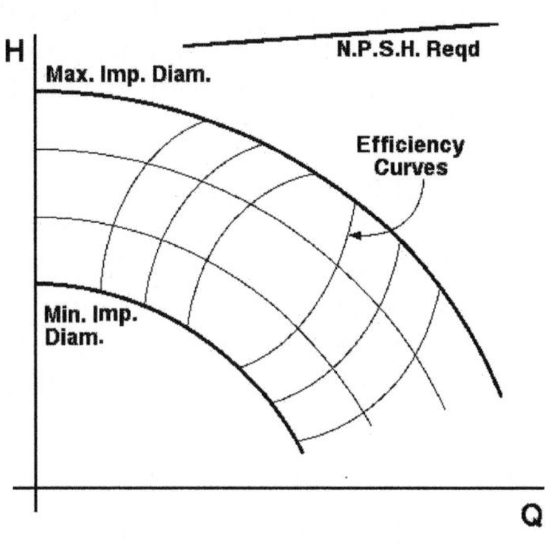

Figure 2.8: Composite head-capacity curve

The highest curve in this series will represent the maximum diameter while the lowest curve will represent the smallest possible diameter of that impeller. Below this minimum diameter, the impeller does not function properly. The intermediate curves represent impeller diameters somewhere between maximum and minimum and are usually selected arbitrarily for reference purposes only.

On a Composite Curve, the points of common values of efficiency are identified on each head-capacity curve. They are then connected to produce a new set of curves, which identified the Pump Efficiency. The curve showing the NPSH Required can also be included.

If the pump style, whose performance is being depicted on the composite curve, is frequently used on cold water, the Brake Horse-power draw can also be displayed as shown. When this happens, the BHP shown will be calculated based on handling a liquid with a specific gravity of 1.0 only.

When an anticipated pump performance is identified on the Characteristic Pump Performance Curve at (for example) 500 gpm at 60 feet total dynamic head, it should be noted that the 500 gpm will only be achieved when the differential pressure across the pump is at 60 feet of head. If the pressure in the system changes at any time, causing a change in the total dynamic head across the pump, the flow rate will also change accordingly. As the total dynamic head increases, the flow will decrease, and as the total dynamic head decreases, the flow will increase. Further details on this topic are discussed in Chapter 3.

2.2.3.1 The best efficiency point (BEP)

It has already been stated that the Best Efficiency Point (BEP) is the most stable condition at which the pump can operate. Therefore, in order to achieve the highest degree of reliability possible for the pump, it should operate as close as possible to the BEP. If the operating flow moves away from the BEP, that reliability decreases. Consequently most pump users will attempt to operate their pumps within a range of 70% to 120% of the BEP. Depending on the hydraulic design of the pump and the service involved, it may be necessary to operate even closer to the BEP.

While the pump can still operate outside this range, it does so at the expense of the reliability of the seals and bearings, as a number of other conditions start to be a factor which will detrimentally impact these items. Other parts of the pump, such as the impeller, volute and shaft will also be subjected to the adverse operating conditions which can contribute increased erosion and fatigue impact to these items.

2.2.3.2 Pump run-out

As every centrifugal pump does not operate reliably beyond a certain flow rate, the published performance curve for that pump is discontinued at that point. This is referred to as the Run-Out condition. Operation of the pump beyond that point (and often, even approaching that point) will cause damage to the pump and will also frequently overload the motor driver.

The damage caused at high flow rates will frequently be a result of cavitation as the increase in flow rate through a pump requires a much higher Net Positive Suction Head. Further details on this matter will be discussed in Chapter 4.

2.2.3.3 Minimum flow point

On every pump curve, a number of 'minimum' flow points can be identified, depending on the operating requirements and equipment reliability standards of the individual end user.

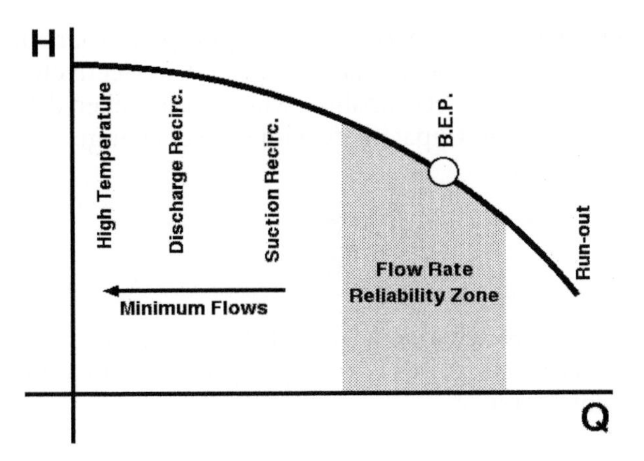

Figure 2.9: Conditions at various flows

The earliest minimum flow point used was that point on the curve at which the flow was so low that it would result in a significant temperature increase of the pumpage. As we have since identified a number of other concerns which relate to low flow conditions, the temperature increase has faded into insignificance.

Suction Recirculation is a condition with similar symptoms to Cavitation that will be discussed more fully in Chapter 4.6. It is a condition created by low flow operation and it frequently dictates the low flow limit of stable operation discussed in Chapter 2.2.3.1 above, in relation to the percentage of BEP. In some industries it is referred to as the 'Minimum Flow for Stable Operating Condition'.

Discharge Recirculation is another condition precipitated by low flow operation that takes effect at a lower flow than Suction Recirculation, and also displays similar symptoms.

2.3 Affinity laws

The Head and Capacity produced by a centrifugal pump is dependent on the velocity with which the liquid leaves the impeller, and is referred to as the peripheral velocity. Therefore the output of the pump can be adjusted by changing the peripheral velocity. This can be accomplished in two ways, with almost identical results:

- by changing the speed of rotation of the impeller or,
- reducing the diameter of the impeller.

Lowering the rotational speed by 20%, will have a similar effect on the Head and the Capacity as would reducing the impeller diameter by 20%.

The relationships that exist between the pump output and the peripheral speed of the impeller are identified collectively as the 'Affinity Laws'. In one such relationship, when an impeller diameter is changed from D_1 to D_2, the Capacity developed will change from Q_1 to Q_2, in the same proportion.

$$\frac{D_2}{D_1} = \frac{Q_2}{Q_1}$$

However, the same relative change in impeller diameter from D_1 to D_2, will change the Total Head from H_1 to H_2 in proportion to the square of the variation in diameter.

$$\left(\frac{D_2}{D_1}\right)^2 = \frac{H_2}{H_1}$$

It must be stressed that this relationship only applies to different diameters of the same impeller, and not to different impellers.

Exactly the same relationships apply when the pump rotational speed changes instead of the impeller diameter.

$$\frac{Q_2}{Q_1} = \frac{D_2}{D_1} \text{ or } \frac{N_2}{N_1}$$

Thus, a change in Capacity from Q_1 to Q_2, can be achieved by, either changing the impeller diameter, or by changing the pump speed by the same ratio.

Similarly a change in Total Head from H_1 to H_2, can be achieved by changing the impeller diameter or by changing the pump speed by the same ratio.

$$\frac{H_2}{H_1} = \left(\frac{D_2}{D_1}\right)^2 \text{ or } \left(\frac{N_2}{N_1}\right)^2$$

It is important to note that a change in impeller diameter or speed affects the whole performance range of the pump, and not just one specific point.

Any reduction in peripheral speed, however caused, will make it appear as though the performance curve will shrink down to the left on the graph. Conversely, a larger impeller diameter or faster pump speed will show the curve extending up to the right.

The relationship of the Power Draw to the rotation speed or impeller diameter may be estimated as being in the same proportion as the change in speed or diameter raised to the third power.

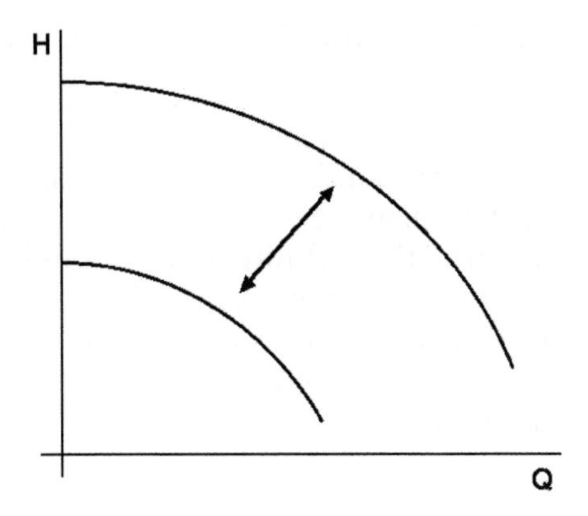

Figure 2.10: Effect of speed or diameter change

$$\frac{HP_2}{HP_1} = \left(\frac{D_2}{D_1}\right)^3 \text{ or } \left(\frac{N_2}{N_1}\right)^3$$

It must be noted however, that this is only an approximation as the pump efficiency may change with the change in peripheral velocity, however it may have been accomplished.

2.3.1 Change of electric motor for performance change

As a change of an electric motor to a higher speed unit is not an uncommon way to effect an increase in pump output, please take very careful note of the results of this example. Consider a process pump being driven by a 25 HP, 1800 rpm. motor, and the motor is replaced by one running at 3600 rpm.

$$\frac{HP_2}{25} = \left(\frac{3600}{1800}\right)^3$$

From the Affinity Laws we can calculate that we would have to replace the 25 HP, 1800 rpm motor with a 200 HP motor running at 3600 rpm. Unfortunately, most pumps purchased to operate on 25 HP are unlikely to be strong enough to accept 200 HP on the same shaft and bearings, thus resulting in premature failure.

2.3.2 Other changes

A change in the NPSH Required can also be estimated by the Affinity Laws, but only when the pump speed is changed. Any change brought about by a change in impeller diameter does not need the Affinity

Laws, as the new NPSH Required can usually be read from the same pump curve.

With these Affinity laws, it becomes quite straightforward to calculate the change in Performance Conditions which results from a change in pump speed or impeller diameter. It should be noted that, for any change greater than 10%, the Affinity Laws should be considered an approximation. Although an approximation is all that is often required, if a set of pump performance curves is available at the alternative speed, these should always be used in preference to the Affinity Laws as they will be much more accurate.

2.4 Pump performance on special liquids

A number of liquid types will detrimentally affect the performance of a centrifugal pump, in that they reduce the head with some reduction in capacity. More importantly, they will lower the efficiency with a resulting marked increase in the horsepower draw needed to drive the pump.

The performance curve shown in Figure 2.11 is a general depiction of

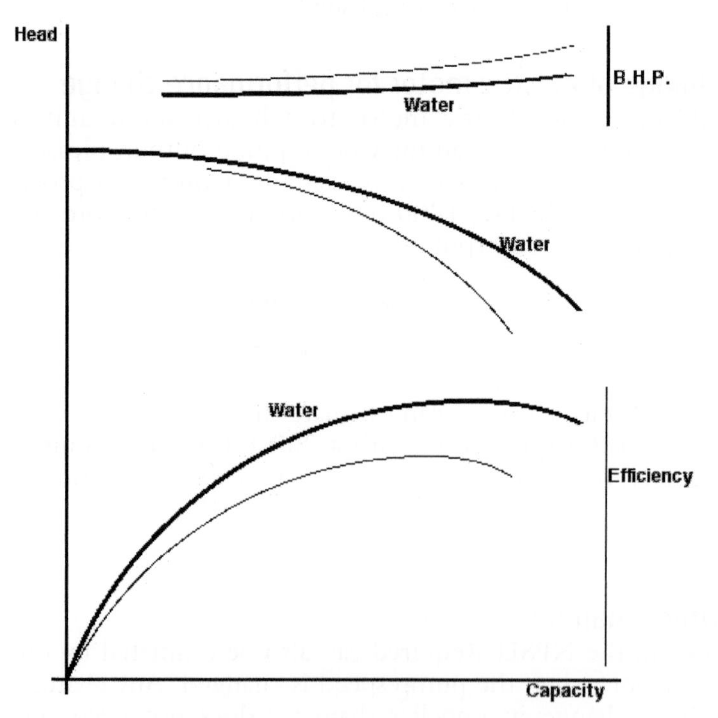

Figure 2.11: Effect of special liquids

the difference between operating the pump on a water service, and the operation of the same pump on either a viscous liquid or on paper stock. With viscous liquids, the performance gradually drops off to a point where positive displacement pumps have to be introduced.

On paper stock with a consistency up to 3%, the pump will perform as though it's handling water. Between 3% and about 6%, the adjustments shown will come into effect. Above 6% consistency the centrifugal pump may require some modifications depending on the fluidity of the stock and the ability of the system to deliver it freely to the impeller. With higher consistencies, a positive displacement screw pump is usually used.

Further details on these and on slurry applications (which can also include a viscous component) can be found in Chapter 8.

2.5 Impeller hydraulic loads

In an earlier section of this chapter, we identified that the Best Efficiency Point (BEP) is the most stable operating condition for that pump and is a direct result of the criteria used in that pump's design. These criteria include the hydraulic loads that act on the pump impeller. As they more directly relate to the pump bearings, the axial hydraulic loads will be discussed in Chapter 7.2.

The radial hydraulic loads however, act around the impeller at right angles to the shaft. Single volute pumps are designed in such a way as to balance out these radial hydraulic loads as much as possible, but a resultant hydraulic force will impact the impeller on a plane at 60° from the cut-water, as shown in Figure 2.12. When the pump operates at the best efficiency point, this force is at a minimum. However, when the pump operation moves away from the BEP, the balance of the hydraulic loads is increasingly compromised and the resultant force can increase dramatically.

In larger process pumps, where the impeller diameters are in excess of 13 inches, the radial forces are balanced out by means of a double volute casing design. This essentially creates an equal and opposite action of all the radial forces around the impeller.

Figure 2.12: Radial forces in single volute casing

This arrangement is not frequently used for pumps designed to handle solids, where the second volute would cause considerable clogging problems. However, there are a few exceptions, and their success tends to depend on the nature of the slurry and the specific design of the casing.

Figure 2.13: Radial forces in double volute casing

2.5.1 Radial thrust

The magnitude of the radial thrust at Shut-off condition (zero flow) will depend on the design of the impeller and the shut-off head. A radial thrust factor (Kso) can be established and tends to vary between 0.15 and 0.38 depending on the design of the impeller and its specific speed. This factor is then used in the equation shown below to calculate the approximate radial force (Fso) that can be expected at the shut-off condition.

$$Fso = Kso \times Pso \times D \times B$$

where
- Kso = Radial Thrust Factor
- Pso = Differential Pressure at Shut-off
- D = Impeller Diameter
- B = Impeller Width at perimeter (incl. shrouds)

In a fairly typical process pump where the impeller is 13 inches in diameter and the operating speed is 3600 rpm, the radial force can be as high as 800 pounds at the shut-off condition. Radial force values at other operating conditions can be approximated by the following equation when the test data exponent (x) is available.

$$F = Fso \left[1 - \left(\frac{Qn}{Q} \right)^{x} \right]$$

where
- Fso = Radial Force at Shut-off
- x = Exponent based on pump test data
- Q = Capacity at Operating Condition (USGPM)
- Qn = Capacity at BEP in USGPM

From the above equation, it can be readily identified that the radial force is at its maximum at the shut-off condition when Q = 0. As the flow rate (Q) increases, the radial force decreases to a theoretical zero at the B.E.P. When the flow exceeds the B.E.P., the radial force will correspondingly increase, but as a negative value. This indicates that the force is now acting in the opposite direction from that indicated in Figure 2.12.

In the absence of any test data and, as a rough estimate only, the value of the exponent (x) may be assumed to vary linearly between 0.7 at an impeller specific speed of 500, and a value of 3.3 at an impeller specific speed of 3,500. In a typical process pump it can be shown that the impeller radial force developed at a capacity halfway between Shut-off and the best efficiency point can be as high as 600 pounds.

2.5.2 Shaft deflection

Having estimated the radial force, it is now possible to calculate the shaft deflection from the formula shown below.

$$y = \frac{F}{3E}\left[\frac{L^3 - A^3}{I_L} + \frac{A^3}{I_a} + \frac{L^2 S}{I_s} + \frac{3X}{2}\left(\frac{L^2 - A^2}{I_L} + \frac{A^2}{I_a} + \frac{2LS}{3I_s}\right) + \frac{X^3}{2I_a}\right]$$

where Moment of Inertia I = $\pi \times d4/64$

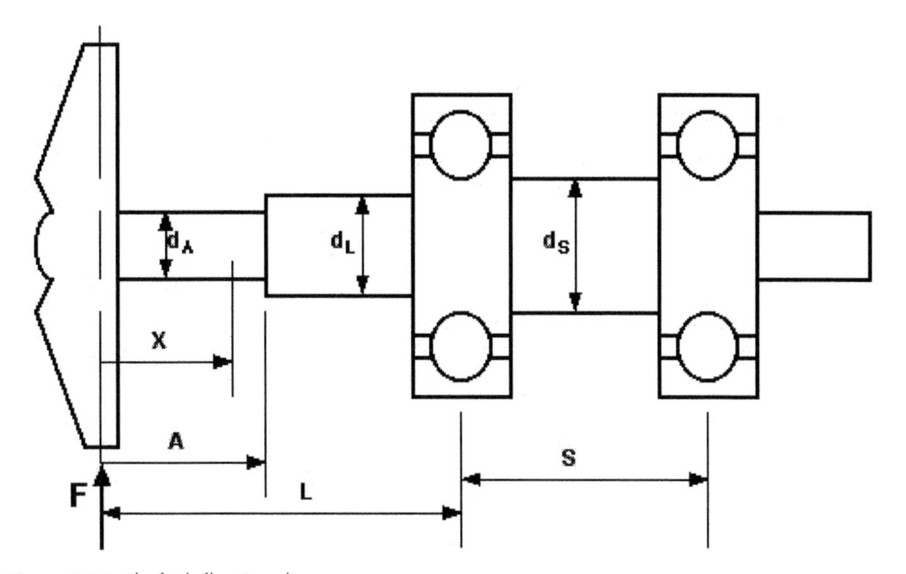

Figure 2.14: Shaft deflection diagram

From this equation, the actual shaft deflection (y) can be calculated at the point on the shaft measured at dimension 'x' from the impeller centerline. This point represents the operational location of the seal faces which is generally considered to be at the face of the stuffing box. At this point many industry standards require the deflection of the shaft to be no more than 0.002 inches (0.5 mm).

Owing to the potential of these high unbalanced hydraulic influences, the capability of the pump shaft to handle such loads without excessive deflection becomes very important. Troubleshooting details of such a condition can be found in Chapter 11.4.2.2, using the simplified version referred to as the Slenderness Ratio.

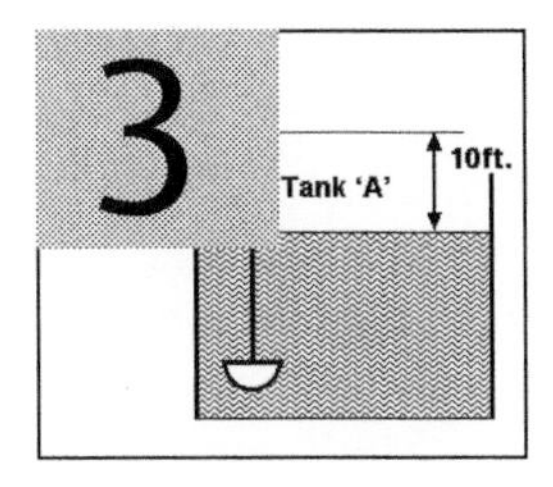

System Hydraulics

3.1 Pump limitations

A centrifugal pump is designed and produced to supply a whole range of head-capacity conditions as identified by it's performance curve. The pump will operate on that curve if it is driven at the particular speed for which the curve is drawn. However, the actual conditions on that curve at which the pump will run, will be determined by the system in which it operates. So, for all practical purposes, the system controls the pump and will determine the conditions at which the pump will operate, regardless of the Head and Capacity for which it was purchased.

This is a considerable advantage from the safety aspect of the System, in that the Centrifugal Pump is not normally capable of over-pressurizing the System. However, in order to understand how the Centrifugal Pump operates in a System, it is first necessary to understand some aspects of system hydraulics and some of the more rudimentary considerations of system design.

3.2 Liquid flow in pipes

For those who may be unsure of the manner in which liquids actually respond to flowing through pipes, the following basic guidelines are offered.

3.2.1 Specific gravity

Specific Gravity (S.G.) is used frequently in the discussion of fluids. It is the name given to the ratio of the density of a liquid to the density of cold water (specifically at 60°F). Therefore, when dealing with cold water, its value of S.G. is 1.0.

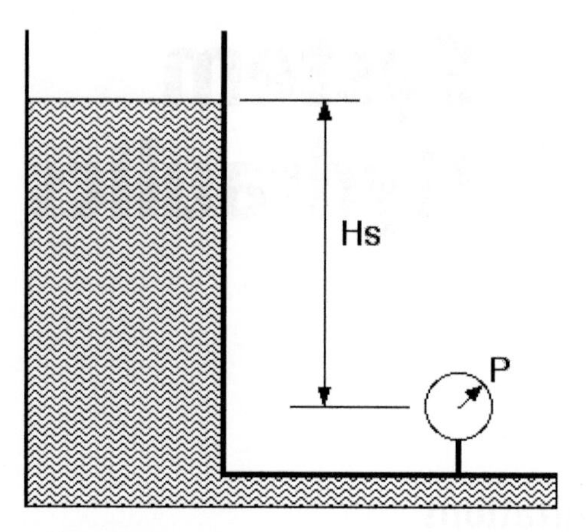

Figure 3.1: Pressure in static system

3.2.2 Pressure in a static system

When a body of liquid is at rest in a system, the relationship between the pressure showing on the gauge and the depth of the liquid above it, will be as shown.

$$P = \frac{Hs \times s.g.}{2.31}$$

where P = Gauge pressure in pounds per square inch

Hs = Static Head of liquid in feet

s.g. = Specific Gravity of the liquid being pumped

3.2.3 Pressure in a flowing system

When a body of liquid is moving in a system, the pressure will drop, as some of the energy supplied by the Static Head is now being lost to Friction. Therefore, even when we maintain the level of water in the tank as shown in Figure 3.2, to stabilize the Static Head, the pressure reading on the gauge will be less than when the liquid in the system was not flowing.

$$P = (Hs - Hf) \times \frac{s.g.}{2.31}$$

where P = Gauge pressure in pounds per square inch

Hs = Static Head of liquid in feet

Hf = Friction Head in feet

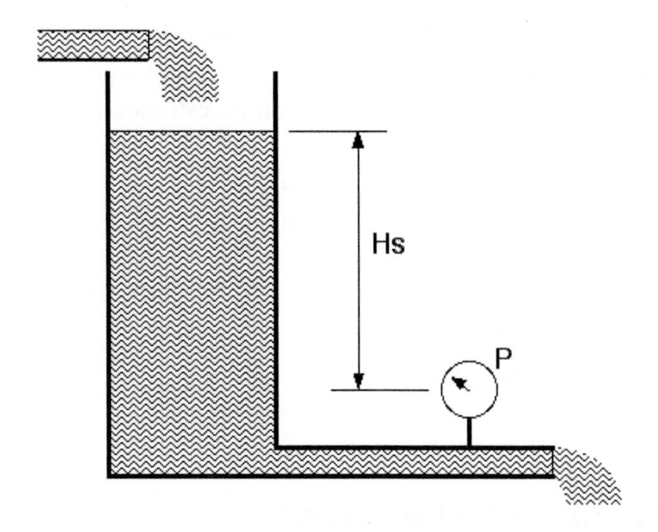

Figure 3.2: Pressure in a flowing system

3.2.4 Changes in an existing system

Please note that the following relationships are offered for those who wish to approximate the effect of changes in an existing system. For accurate data, it is recommended that the Hydraulic Institutes' Friction Loss tables are consulted, either in their Engineering Data Book, or as shown in Chapter 14 of this book.

3.2.4.1 The effect of capacity change on friction

When the flow rate is changed without changing the pipe size (i.e. $D_2 = D_1$), the approximate change in Friction Loss can be estimated as shown.

$$\left[\frac{Q_2}{Q_1}\right]^2 = \frac{Hf_2}{Hf_1}$$

where Q = Flow Rate

Hf = Friction Loss

In other words, the Friction Loss will vary as the square of the Flow Rate.

3.2.4.2 The effect of head change on flow rate

When the Static Head is changed, again without changing the pipe size, the approximate change in Flow Rate can be estimated as shown.

$$\left[\frac{Q_2}{Q_1}\right]^2 = \frac{Hs_2}{Hs_1}$$

where Q = Flow Rate

Hs = Static Head

3.2.5 Pipe size changes in a system

The change in pipe diameter may be necessary to reduce friction losses, or increase the NPSH available to the pump (see Chapter 4.4.2). Therefore, when capacity is unchanged ($Q_2 = Q_1$), the friction loss is in INVERSE proportion to the 5th power of the change in the pipe diameter, as shown below.

$$\left[\frac{D_1}{D_2}\right]^5 = \frac{Hf_2}{Hf_1}$$

where D = Pipe Diameter

Hf = Friction Loss

To operate with the same head (i.e. $H_2 = H_1$), such as from a lake or river, with the new pipe size, the following approximation will apply.

$$\left[\frac{D_2}{D_1}\right]^{2.5} = \frac{Q_2}{Q_1}$$

where D = Pipe Diameter

Q = Flow Rate

3.3 Basic elements of pump system design

In designing any kind of pumping system, the first requirement is to determine the speed at which the task must be performed. In other words, the flow needed through the system. In some systems, the flow rate will be determined by production requirements or by other process considerations such as the flow rate needed to achieve the necessary temperature transfer in a liquid flowing through a heat exchanger. For the sake of this exercise, let us consider a batch process system where the average flow rate can be calculated by dividing the volume to be transferred, by the time allowed for that transfer.

The next requirement to be considered is how to overcome all the factors which hinder the movement of the liquid from one point to another in the system. These are primarily Gravity and Friction and we will deal with them separately.

3.3.1 Gravity and static head

If we consider Gravity as a force of nature that drives vertically downwards then, in a pumping system, we can oppose it by means of an energy factor we will refer to as the Total Static Head. This is simply the change in elevation through which the liquid must be lifted, and is measured vertically, regardless of the linear distance between the start and end points in the system. As shown in Figure 3.3, the Static Head

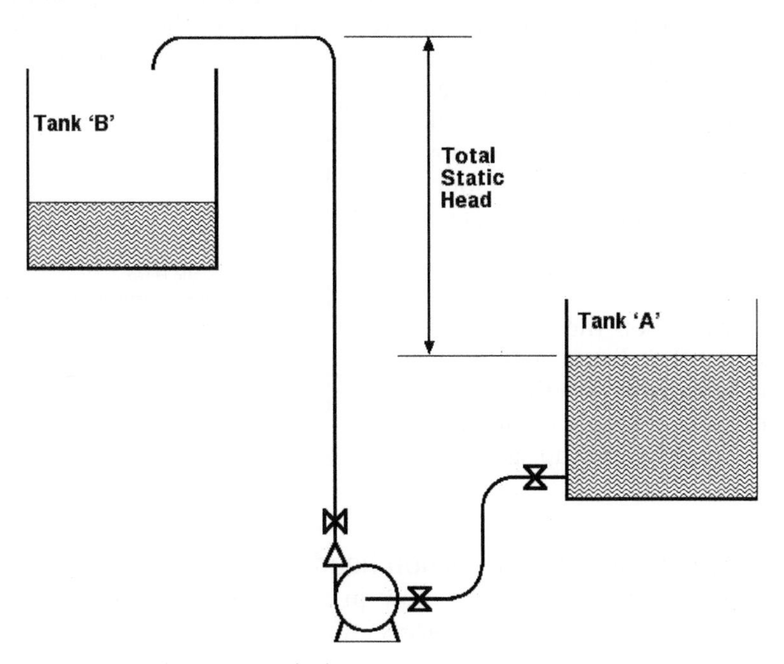

Figure 3.3: Static head to high point in the line

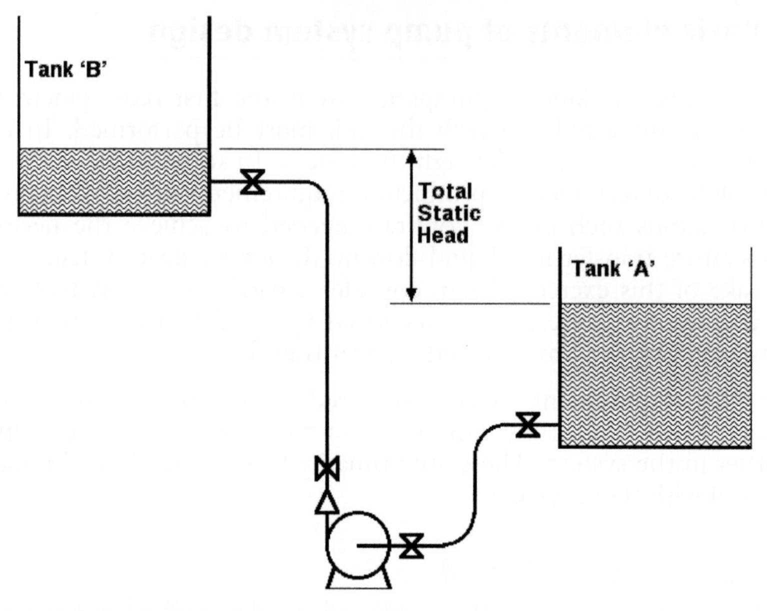

Figure 3.4: Static head entering bottom of the tank

can be measured between the free surface of the liquid in Tank 'A', and the highest level of the discharge line.

In another system (Figure 3.4), where the pump discharges into Tank 'B' at a point below the liquid level in that tank, the Total Static Head in the system is the vertical distance from the free surface of the liquid in Tank 'A' to the free surface of the liquid in Tank 'B'. It should be noted that this also applies, even when the suction source is lower than the pump. This will be discussed in greater detail in Chapter 4.

Regardless of the layout of the system, the Total Static Head is not a variable of the flow rate, therefore a graph comparing the two would show up as a straight line.

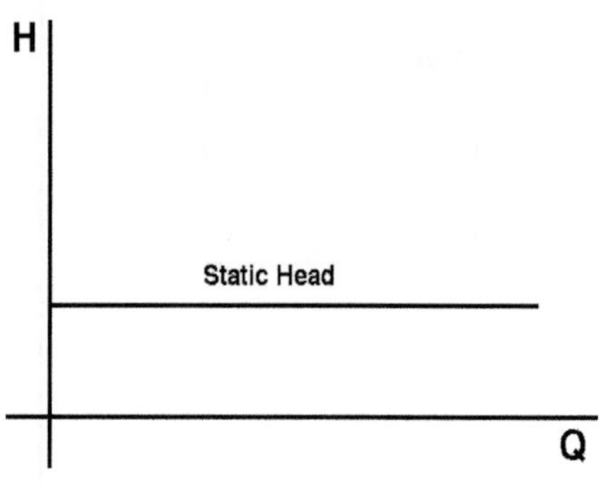

Figure 3.5: Static head curve

3.3.2 Friction losses and friction head

Friction is the resistance to flow in the piping system and must be considered for three separate areas individually.

- the Piping
- the Valves and Fittings, and
- Other Equipment, such as filters, heat exchangers, etc.

The Friction Losses in piping can most readily be obtained from the Friction Loss Tables available from a variety of sources such as the Standards of the Hydraulic Institute. For benefit of the reader, many of these tables are reproduced in Chapter 14 of this book. Tables are also available to identify the losses through the more common pipe fittings and valve types. However, any such losses in Filters, Heat Exchangers, etc., must be obtained from the original equipment manufacturer, or by measuring the equipment on site.

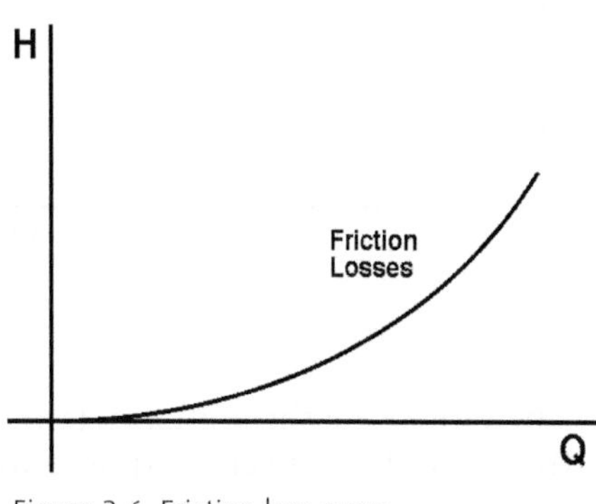

Figure 3.6: Friction loss curve

As the flow increases, so too does the Friction Loss but at a far higher rate as shown in Figure 3.6.

3.3.3 Velocity head

Another factor that has to be overcome is the head required to accelerate the flow of liquid through the pump. This is the difference in the values of Velocity Head ($V^2/2g$) at the Suction and Discharge Nozzles of the pump.

As the linear velocity of the liquid in most systems is maintained at lower than 10 ft./sec. (3 m/sec.), the Velocity Head is usually an insignificant part of the total, except in low head applications.

3.3.4 Total head

The combination of these values equals the Total Head of the System.

Total Head = Static Head + Friction Loss + Velocity Head

3.4 System curve

When the Total Head (H) is plotted against the Flow Rate (Q), the resultant curve is known as the System Curve (Figure 3.7).

Therefore, when a specific Flow Rate is selected for a system, the System Curve will identify the Total Head that must be overcome.

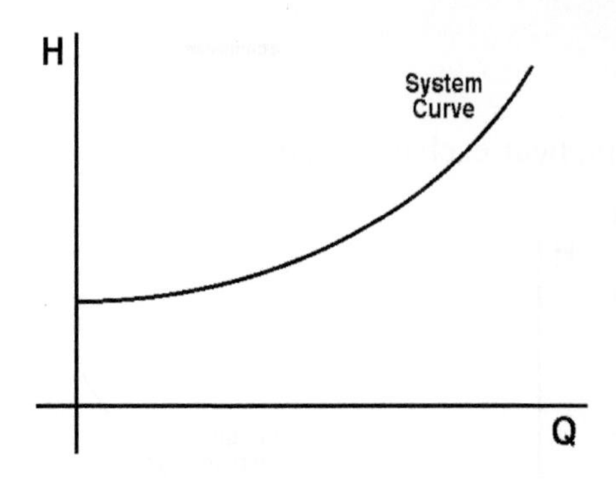

Figure 3.7: System curve

Figure 3.8: Pump/system curves

The Flow Rate through a system can only be supplied by a pump, and is therefore the Capacity required from the pump.

The intersection of the Pump Performance Curve and the System Curve represents the point at which the pump will operate as shown in Figure 3.8.

In systems where the flow rate is maintained at a constant level, the conditions identified in Figure 3.8 will not change. In other systems however, the operating condition of the centrifugal pump is constantly changing.

3.5 The effect of operating performance

3.5.1 Static head changes

3.5.1.1 Batch transfer system

In a batch transfer system for example, the most obvious change is that which is created by emptying the supply tank. This results in a reduction in the liquid level in that tank and the equivalent increase in the static head that the pump must overcome.

When this happens, the System Curve will move straight up on the graph, with the following three specific conditions occurring during a single batch.

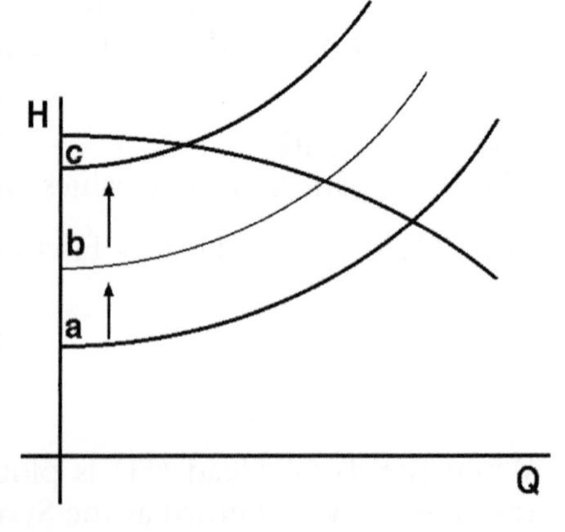

Figure 3.9: Varying system curves

At the Startup point, the level of liquid in the supply tank will be at its highest, while the level in the discharge tank could be zero. This will translate into a low value of Static Head (Curve a).

At the Intermediate point, the level of liquid in the supply tank will have dropped, and the level in the discharge tank will be greater. This will result in a higher value of Static Head (Curve b).

At the Shutdown point, the liquid in the supply tank will have been transferred entirely into the discharge tank, resulting in the maximum value of Static Head (Curve c). At this point, the pump should be shut down.

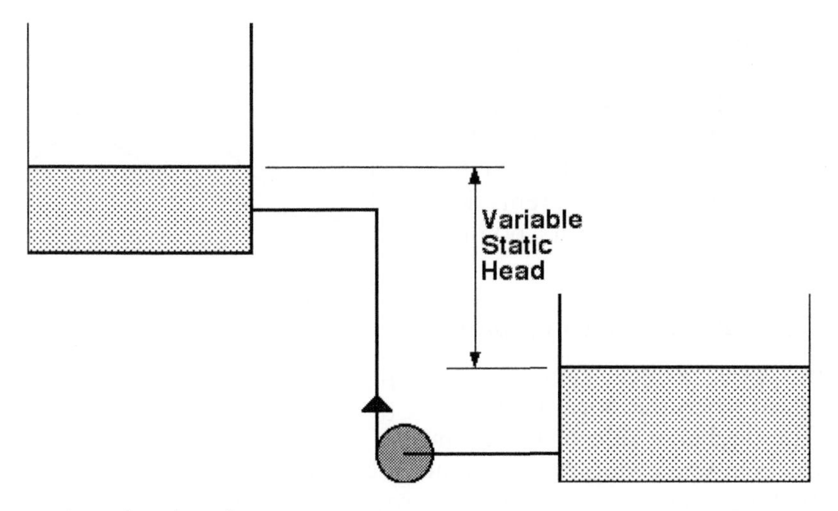

Figure 3.10: Batch system diagram

The system curve will assume the three positions shown in Figure 3.9 as it moves steadily from Startup (a) to Shutdown (c) with the corresponding change in pump capacity. However, as the system approaches the Shutdown point, the pump performance will become unstable. This is due to gradual elimination of appropriate suction conditions which will be discussed in some detail in Chapter 4.

3.5.1.2 Pressurized system

In a pressurized type of system, such as a boiler feed system, the feed pump takes it's suction from a deaerator under vacuum and supplies a boiler under pressure. In this system, the Differential Pressure is not a function of the flow rate and will have similar consequences as the Static Head. Any change in pressure in either the deaerator or the boiler, will also cause the system curve to move up or down as indicated in Figure 3.9.

3.5.1.3 Closed loop system

A closed loop system is one in which the entire system is pressurized by the pump. To achieve this, the pumpage is fully contained within a series of pipes and pressurized process equipment all the way from the pump discharge, through the system, and back to the pump inlet. In such a layout, the Static Head in the system is effectively zero.

3.5.2 Friction head changes

3.5.2.1 System controls

A change in Friction Loss can be caused by a variety of conditions such as manual operation or automated controls opening and closing a different valving system. This will result in the System Curve adopting a different slope that will pivot about its point of origin at zero capacity.

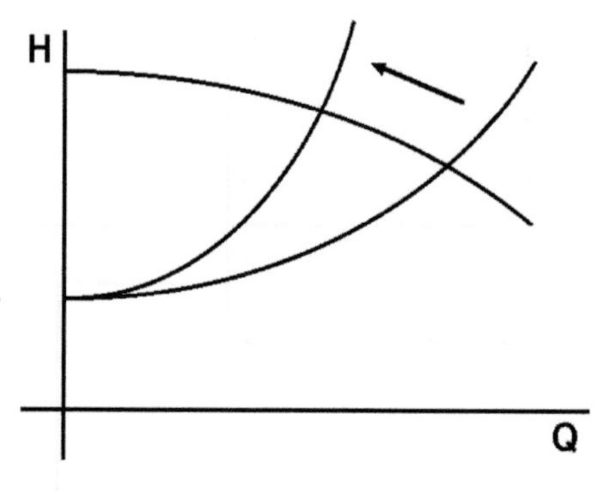

Figure 3.11: System curve for friction changes

3.5.2.2 Restrictions in piping

The same effect can also be realized when the bore of the pipe in the discharge side of the pump reduces in size owing to some kind of buildup such as scaling, etc. Such a buildup may also occur inside process equipment such as filters or heat exchangers. These build-ups automatically reduce the bore of the pipe and therefore increase the Friction Losses in that pipe.

It is worthwhile to compare the difference in effect between throttling a valve and having scale build up in the pipe. The former will have an immediate effect, while the scale build-up can frequently take years to show some effect.

3.5.3 Recirculation systems

When a pump is required to operate under two distinctly different operating conditions and the flow needs to be cut back occasionally to a considerable extent, a recirculation line should be used. This allows the system to continue to operate without loss of pump efficiency. It redirects a percentage of the flow back to the suction source through valves that are specially designed for this purpose.

One of the most important factors in a recirculation system is to ensure that the liquid is recirculated back to the suction source of the pump. It should not be piped back to the suction line immediately preceding the

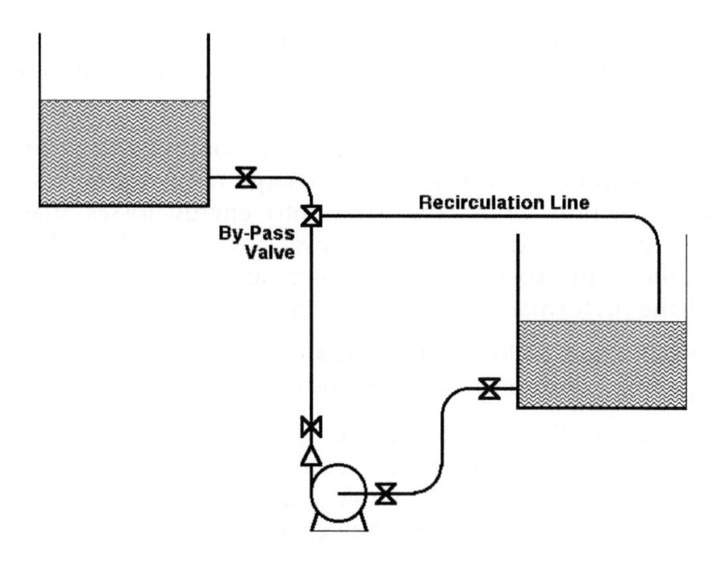

Figure 3.12: Recirculation system

pump suction as this will cause considerable turbulence at the inlet to the pump, resulting in reduced operating performance.

Even when being delivered back to the suction source, it should be piped up in such a way to ensure the minimum amount of turbulence in the suction tank. This may be a simple matter of introducing the recirculation into the other side of the tank from the location of the pump suction outlet. It may even be necessary to introduce some kind of baffling plate to block the turbulence from flowing to the pump suction outlet.

3.5.4 Pump speed changes

If a change in Flow Rate is noted during operation, it usually means that the system has been changed either manually, or by an automated control system, such as is indicated in Figure 3.11. The only exception to this would be the case where the pump is driven by a variable speed mechanism and the pump curve therefore moves on the system curve.

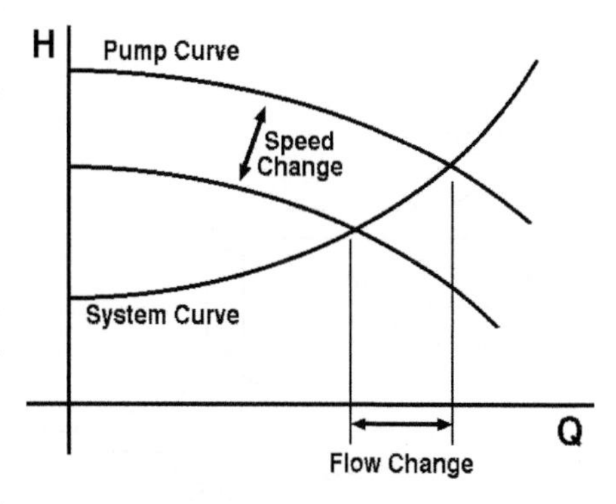

Figure 3.13: Pump speed change curve

Variable Speed Adjustment is an efficient method of modifying the performance of a pump and system, and can be achieved by a variety of mechanical or electrical drives. The Variable Frequency Drive is one of the most commonly used items in many plants.

When frequent adjustment is needed to the output of the pump, the traditional method of throttling the discharge valve absorbs a significant amount of friction which translates into energy losses that can be identified in increased heat and excessive wear in the valve. It also restrains the pump to operate at a lower and (usually) a less efficient point on the performance curve, thus compounding the energy loss.

A speed reduction to lower the pump output will frequently have the pump operating with only a very minor reduction in efficiency.

Computer controlled speed change systems are now available that permits the end user to preset the required pumping condition (either flow rate or head) and the Variable Frequency Drive will adjust the pump speed automatically to meet all changes in system demand.

3.5.5 Series and parallel operation

In many instances, two or more pumps are required to operate together, either in Series or in Parallel. In a Series Operation, each of the two pumps operate at the same flow rate, but share the head, while in a Parallel Operation, each of the two pumps operate at the same head, but share the flow rate.

3.5.5.1 Series operation

This arrangement is frequently used where a larger pump cannot operate with the NPSH being made available from the system. A smaller pump is therefore installed upstream of the larger one to boost the Suction Pressure to the larger pump.

It is important to note that, under these conditions, the smaller pump

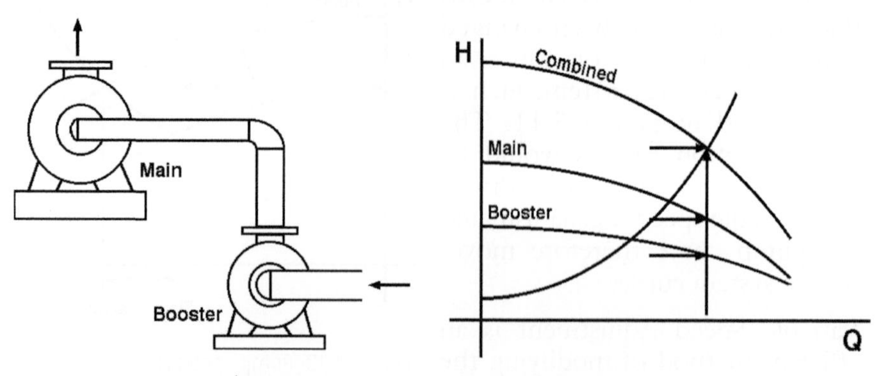

Figure 3.14: Series pump diagram

Figure 3.15: Series pump curve

must be capable of handling the same Flow Rate as the larger pump. Only the Head is being changed.

The ultimate example of Series Operation is the multistage pump where the first impeller pumps into the second and then the third, etcetera. This results in a high pressure pump with all the impellers operating at the same capacity.

3.5.5.2 Parallel operation

In the more common Parallel Operation, banks of pump are used in parallel where they all take their suction from a common header and discharge into a common header. Each of the two (or more) pumps operates at the same Head, but share the Flow Rate.

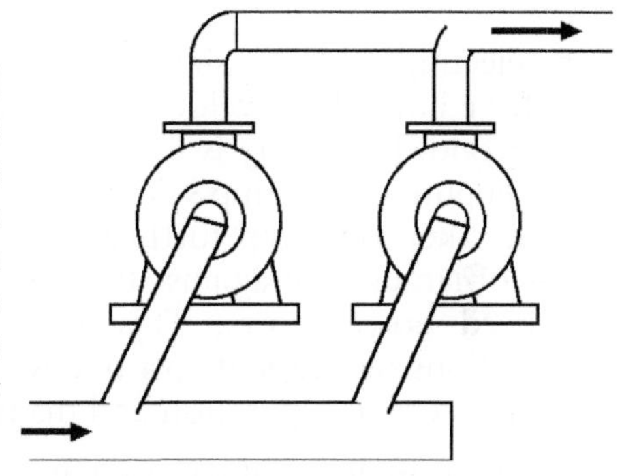

Figure 3.16: Parallel pump diagram

Because of the slope of the System Curve, the pumps in this arrangement will each operate at a lower Flow Rate when operating together, than they would if they operate alone on the same system. This is particularly relevant on multi-pump arrangements and requires careful selection to ensure the most efficient and stable operation.

Many industries use banks of pumps in parallel when they are required to adjust the total flow output beyond the economical capability of one pump. Municipal water distribution systems are a prime example where such flexibility is required.

These systems are also susceptible to one of the dangers of putting too many pumps in parallel on the same system. Because each pump operates at a lower Flow Rate when operating together, than they

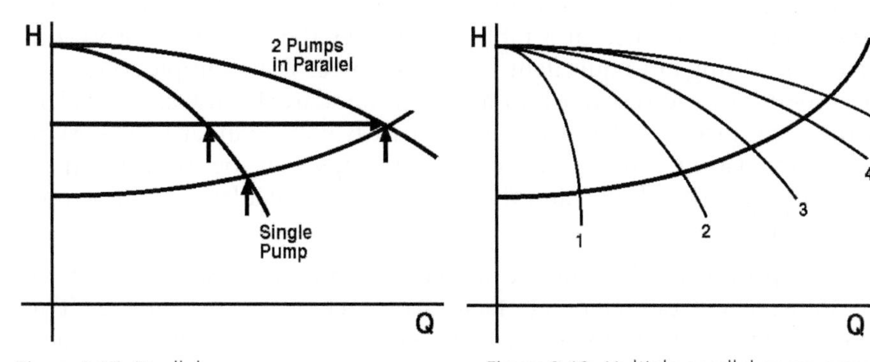

Figure 3.17: Parallel pump curves

Figure 3.18: Multiple parallel pump curves

would if they operate alone on the same system, a steady increase in numbers of pumps will reduce the flow rate through each pump. This could result in the final pump adding only a fraction of its capability to the system output as is indicated in Figure 3.18.

3.6 Pump system analysis

In selecting the pipe sizes to be used in the systems under discussion, we have limited the velocity in the pump discharge line to a value of 10 ft/second. However, as will be discussed in Chapter 5, the introduction of computerized pump selection and system design software, allows this to be taken one step further. It is now possible to balance the higher cost of larger pipe against the lower velocity in the lines that result in reduced power costs. This allows the designer to go beyond the restrictions of a capital cost budget and implement the consideration of lifetime costs of operation and maintenance.

The friction loss values used in these examples are drawn from the Friction Loss tables shown in Chapter 13. The values in these tables are based on the roughness parameter for new Schedule 40 Steel Pipe, with no allowances for age or abnormal conditions of interior surface. Consequently it is a fairly common practice to apply a safety factor to these calculated values, particularly when working with an older system where the interior surface of the pipes may be scaled or rough, or may become so very quickly after start-up of the system. A safety factor is also frequently used if the engineer is working with incomplete information. The amount of any safety factor must be estimated for each installation individually and should be based on local conditions and experience. As the following examples involve fairly short runs of piping and thus have low levels of friction losses, we will use a safety factor of 10% to demonstrate its use.

In order to calculate the friction losses for pipes and fittings, two approaches are possible, depending on the information available. One option requires the use of a table that shows the resistance of various valves and fittings in equivalent lengths of pipe. For example, one such chart identifies the resistance of a 6 inch Standard Elbow as equivalent to the resistance of a length of 16 feet of 6 inch Standard Pipe. With such information, all the valves and fittings can be transferred to equivalent lengths of pipe and the friction losses calculated as shown for straight pipe.

In the approach used in these examples, it is first necessary to establish the Resistance Coefficient (K factor) of each valve and fitting. This information is contained in the charts for Typical Resistance Coefficients for Valves and Fittings in Chapter 14. The value is then multiplied by the Velocity Head (from the Friction Loss tables) to provide the Friction Loss for that particular fitting. Consequently, the value of the friction loss for each fitting must be calculated individually.

3.6.1 Example 1

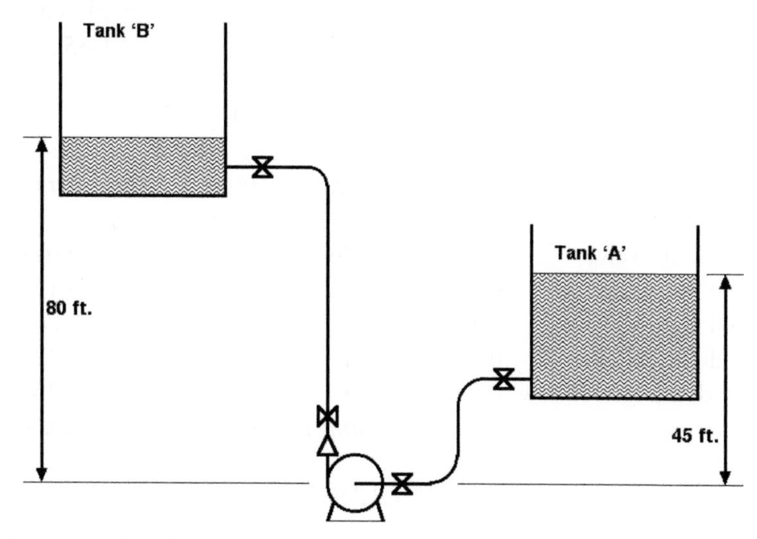

Figure 3.19: Example 1 diagram

Flowrate required through system (as specified)		340 gpm
Maximum acceptable velocity in discharge line		10 ft./sec.
Therefore:	Discharge Line Size (from friction loss tables)	4 inches
	Suction Line Size (one size larger)	6 inches

Calculation of Total Head

$$H = Hs + Hsd + Hv + Hf$$

Static Head (Hs)
The vertical distance from the elevation of the free surface of the liquid in the suction source to the elevation of the free surface of the liquid in the discharge tank.
85 feet Elevation – 45 feet Elevation **Static Head = 40 feet**

System Differential Head (Hsd)
As both tanks are open to atmosphere, **Differential Head = Zero**

Velocity Head at 340 gpm (Hv)
(from Friction Loss Tables)
V2/2g on 4 inch discharge – V2/2g on 6 inch suction
Therefore 1.140 – 0.222 **Velocity Head = 0.918 feet**

Friction Loss (Hf)
Piping Losses based on a Flow Rate of 340 gpm
(from Friction Loss Tables)

Pipe Size	*Length of Pipe*	*Hf per 100 ft*	*Friction Loss*
4 inch	130 feet	6.19 feet	8.047 ft.
6 inch	25 feet	0.806	0.2015 ft.
		Piping Losses	*8.2485 feet*

Valves and Fittings Losses based on a Flow Rate of 340 gpm
(from Resistance Coefficient Tables and Friction Loss Tables)

Item	*K factor*	*V2/2g*	*Hf per fixture*	*Qty*	*Total Hf*
6" flanged Gate Valve	0.11	0.222	0.02442	2	0.04884 ft.
6" flanged Std. Elbow	0.29	0.222	0.06438	2	0.12876 ft.
4" flanged Gate Valve	0.16	1.14	0.1824	2	0.3648 ft.
4" flanged Check Valve	0.2	1.14	0.228	1	0.228 ft.
4" long rad. Flanged Elbow	0.22	1.14	0.2508	1	0.2508 ft.
		Valves and Fitting Losses			*1.0212 feet*

Total Friction Losses = 8.2485 + 1.0212 = 9.2697 ft.

Implementing a 10% Safety Factor = 9.2697 × 1.1

Therefore Total Friction Losses (Hf) = 10.2 feet

H	=	Hs	+	Hsd	+	Hv.	+	Hf
	=	40	+	0	+	0.918	+	10.2
	=	51.118 feet						

Therefore design operating condition = 340 gpm @ 51 feet of Head

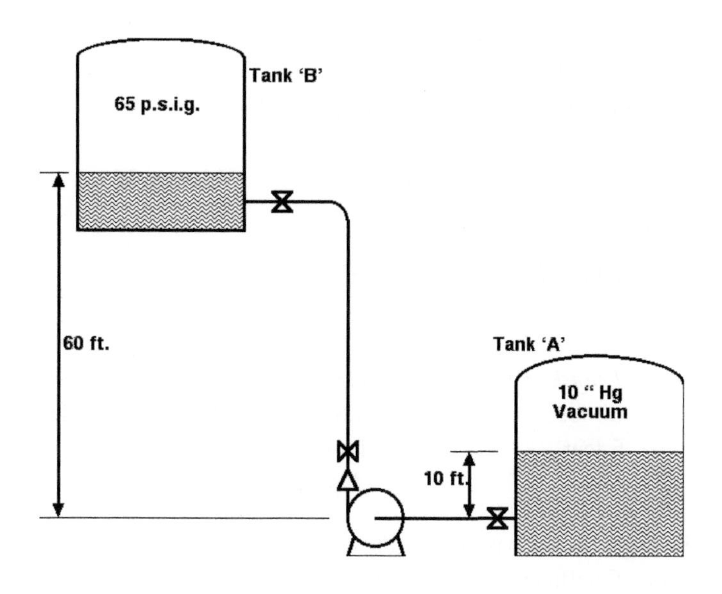

Figure 3.20: Example 2 diagram

3.6.2 Example 2

Flowrate required through system (as specified)		160 gpm
Maximum acceptable velocity in discharge line		10 ft./sec.
Therefore:	Discharge Line Size (from friction loss tables)	3 inches
	Suction Line Size (one size larger)	4 inches

Calculation of Total Head

$$H = Hs + Hsd + Hv + Hf$$

Static Head (Hs)

The vertical distance from the elevation of the free surface of the liquid in the suction source to the elevation of the free surface of the liquid in the discharge tank.

60 feet Elevation – 10 feet Elevation **Static Head = 50 feet**

System Differential Head (Hsd) (from gauge pressures and s.g. = 1.0)

Pressure in Tank 'B'	=	65	×	2.31/s.g.	=	150.15 ft.
Pressure in Tank 'A'	=	10'Hg	×	1.13/s.g.	=	–11.3 ft.
System Differential Head	=	Pressure in Tank 'B' – Pressure in Tank 'A'				
	=	150.15 – (–)11.3				

System Differential Head = 161.45 feet

Velocity Head (Hv) at 160 gpm
(from Friction Loss Tables)
V2/2g on 3 inch discharge – V2/2g on 4 inch suction
Therefore 0.749 feet – 0.253 feet

Velocity Head = 0.496 feet

Friction Loss (Hf)
Piping Losses based on a Flow Rate at 160 gpm
(from Friction Loss Tables)

Pipe Size	*Length of Pipe*	*Hf per 100 ft*	*Friction Loss*
3 inch	80 feet	5.81 feet	4.648 ft.
4 inch	5 feet	1.49 feet	0.0745 ft.
		Piping Losses	*4.7225 ft*

Valves and Fittings Losses based on a Flow Rate of 160 gpm
(from Resistance Coefficient Tables and Friction Loss Tables)

Item	*K factor*	*V2/2g*	*Hf per fixture*	*Qty*	*Total Hf*
4" Flanged Gate Valve	0.16	0.253	0.04048	2	0.08096 ft
3" Flanged Gate Valve	0.25	0.749	0.18725	2	0.3745 ft
3" Flanged Check Valve	0.2	0.749	0.1498	1	0.1498 ft
3" Long Rad. Flanged Elbow	0.26	0.749	0.19474	1	0.19474 ft
	Valves and Fitting Losses				*0.8 feet*

Total Friction Losses = 4.7225 + 0.8 = 5.5225 feet

Implementing a 10% Safety Factor = 5.5225 × 1.1

Therefore Total Friction Losses (Hf) = 6.07475 feet

H = Hs + Hsd + Hv + Hf
50 + 161.45 + 0.496 + 6.075
218 feet

Therefore design operating condition = 170 gpm @ 218 feet of Head

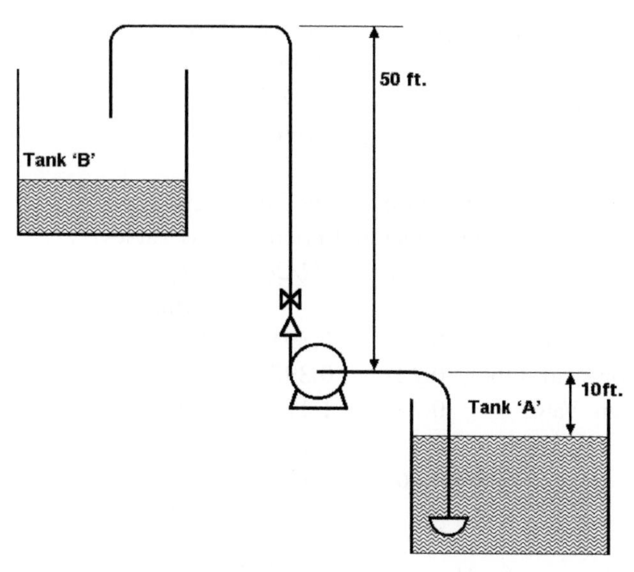

Figure 3.21: Example 3-A diagram

3.6.3 Example 3-A

Flowrate required through system (as specified)		200 gpm
Maximum acceptable velocity in discharge line		10 ft/sec.
Therefore:	Discharge Line Size (from friction loss tables)	3 inches
	Suction Line Size (one size larger)	4 inches

Calculation of Total Head

$$H = Hs + Hsd + Hv + Hf$$

Static Head (Hs)
The vertical distance from the elevation of the free surface of the liquid in the suction source to the elevation of the free surface of the liquid in the discharge tank, or the highest point in the line.
50 feet Elevation – (–) 10 feet Elevation **Static Head = 60 feet**
System Differential Head (Hsd)
As both tanks are open to atmosphere, **Differential Head = Zero**
Velocity Head at 200 gpm (Hv)
(from Friction Loss Tables)
V2/2g on 3 inch discharge – V2/2g on 4 inch suction
Therefore: 1.17 – 0.395 **Velocity Head = 0.775 feet**

Friction Loss (Hf)
Piping Losses based on a Flow Rate at 200 gpm
(from Friction loss Tables)

Pipe Size	*Length of Pipe*	*Hf per 100 ft*	*Friction Loss*
3 inch	70 feet	8.90 feet	6.23 ft.
4 inch	20 feet	2.27 feet	0.454 ft.
		Piping Losses	*6.684 feet*

Valves and Fittings Losses based on a Flow Rate of 200 gpm
(from Resistance Coefficient Tables and Friction Loss Tables)

Item	*K factor*	*V2/2g*	*Hf per fixture*	*Qty*	*Total Hf*
4" flanged Foot Valve	0.8	0.395	0.316	1	0.316 ft.
4" Flanged Long rad. Elbow	0.22	0.395	0.0869	1	0.0869 ft.
3" Flanged Gate Valve	0.25	1.17	0.2925	1	0.29 ft.
3" Flanged Check Valve	0.2	1.17	0.234	1	0.23 ft.
3" Long Rad. Flanged Elbow	0.26	1.17	0.3042	2	0.6084 ft.
		Valves and Fitting Losses			*1.5313 ft.*

Total Friction Losses = *6.684 + 1.5313* = *8.2153 ft.*
Implementing a 10% Safety Factor = 8.2153 × 1.1

Therefore Total Friction Losses (Hf) = 9.0 feet

$$H = Hs + Hsd + Hv + Hf$$
$$= 60 + 0 + 0.775 + 9.0$$
$$= 70 \text{ feet}$$

Therefore design operating condition = 200 gpm @ 70 feet of Head

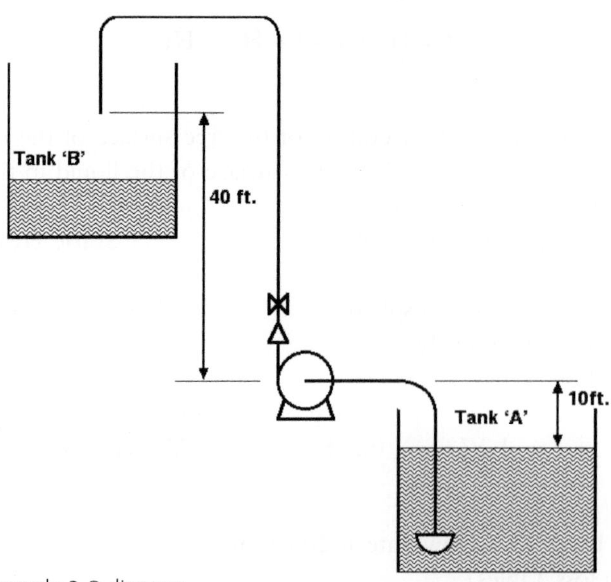

Figure 3.22: Example 3-B diagram

3.6.4 Example 3-B

From the previous Figure 3.21, it can be noted that the static head on the system shown will change from that calculated in Chapter 3.6.3 above, only once the system is fully charged.

The previous calculation used the static head 'to the highest point in the line' and this is necessary in order to ensure that the system is fully charged with liquid. However, once the system is full, a syphon effect will kick in, and this will require the pump to raise the liquid no further than the 40 foot elevation. Consequently, the Total Head can be calculated as follows:

H	=	Hs	+	Hsd	+	Hv	+	Hf
	=	50	+	0	+	0.775	+	9.0
	=	60 feet						

Therefore design operating condition = 200 gpm @ 60 feet of Head

3.6.5 System curve revisited

In each of the above examples, we have only identified the design flow conditions required. By working out the same equations at 50% and 120% of the design flow, we can draw a complete system curve for each example.

In Example 3-A and 3-B, we will have two system curves for the same operation. 3-A will identify the start-up condition, while 3-B will take over on the instant that the high reaches of the piping are filled and the syphon effect takes over.

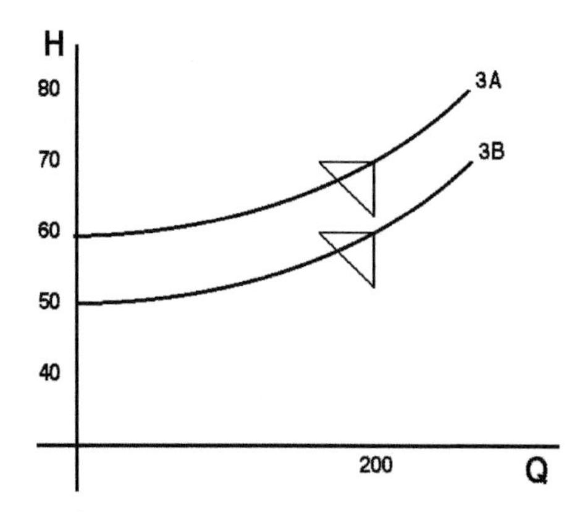

Figure 3.23: System curves for 3A & 3B

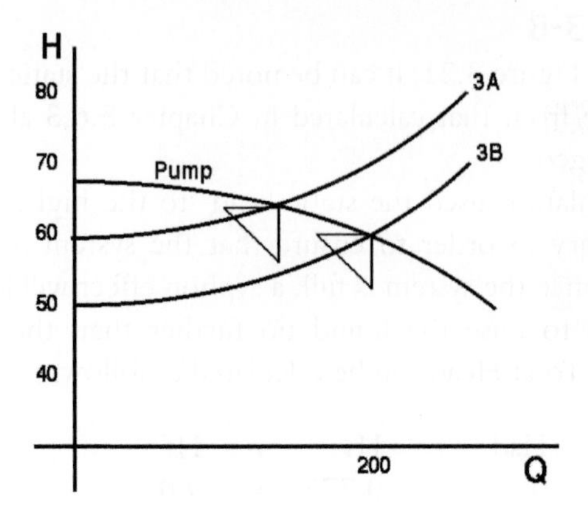

Figure 3.24: System curves for 3A & 3B with pump curve

By selecting a pump for Condition 3B, it will intersect the system curve for Condition 3A at a lower flow rate as shown in Figure 3.24. This simply means that the 'filling' system discussed in Example 3A will take place at less than the originally estimated flow rate, and will take longer to complete prior to the syphoning effect coming into play and the pump moving into normal operating mode.

4 Suction conditions

4.1 General

The suction conditions have to be considered the key element of a successful pump installation and operation. If the liquid to be pumped does not arrive at the impeller eye under the right set of conditions, the pump will be unable to provide the performance for which it was designed. Consequently an understanding of the necessary suction conditions is necessary, as well as an overview of some pump design considerations that affect these conditions.

The most predominant of all suction problems is Cavitation. More paragraphs have been penned on this topic than on every other aspect of pumping combined, yet the vast majority of the world's pumps have never experienced the problem. However, there are enough of those who have been subjected to cavitation for us to review the matter in some detail.

4.2 Vapor pressure

Cavitation is particularly related to a condition referred to as vapor pressure which is that pressure below which a liquid will vaporize. For example, water at 212° F. will vaporize when the pressure falls to 14.7 p.s.i. The layman's term for this phenomenon is 'Boiling'. Similarly, water at only 100° F. will boil or vaporize if exposed to a vacuum of 18 inches of mercury. (18" Hg)

4.3 Cavitation

To anyone who works with pumps, the symptoms of cavitation are relatively familiar. They are a unique rumbling/rattling noise, and high vibration levels. Closer inspection will also reveal pitting damage to the impeller and a slight reduction in the Total Head being developed by the pump. In order to consistently avoid or cure these problems, it is important to understand what cavitation really is and what causes it in a centrifugal pump.

Cavitation is a two part process caused by the changes in pressure as the liquid moves through the impeller. As the liquid enters the suction nozzle of the pump and progresses through, there are a number of pressure changes that take place as shown in Figure 4.1.

As the liquid enters the pump through the suction nozzle, the pressure drops slightly. The amount of reduction will depend on the geometry of that section of the particular pump and will vary from pump to pump. The liquid then moves into the eye of the rotating impeller where an even more significant drop in pressure occurs.

The first part of the cavitation process occurs if the pressure falls below the liquid's vapor pressure in the eye of the impeller. This causes vapor bubbles to be created in that area (in other words, the liquid boils!). The second part of the process occurs as the centrifugal action of the impeller moves the bubbles onto the vanes where they are instantly re-pressurized and thus collapsed in a series of implosions.

While a single such implosion would be insignificant, their increasing repetition and severity develops energy levels well beyond the Yield Strength of most impeller materials. At this stage, the impeller starts to disintegrate and small cavities are created in the metal. This condition also creates the noise and high vibration levels mentioned earlier.

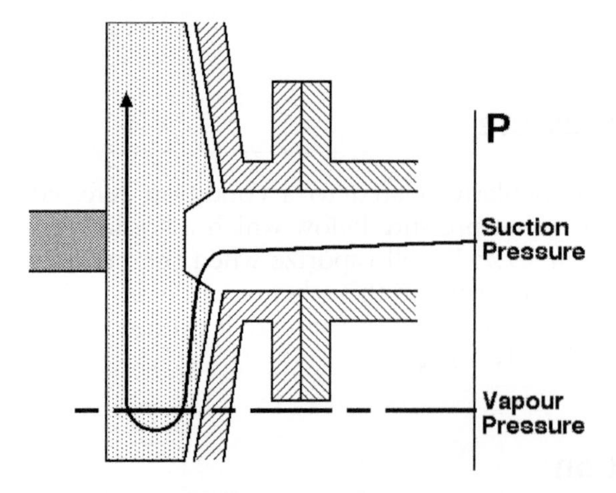

Figure 4.1: Pressure gradient in suction/impeller

When considering Figure 4.1, it is evident that the problem results from the pressure of the liquid dropping below its vapor pressure in the eye of the impeller. This is what creates the vapor bubbles in that area. Consequently, cavitation can usually be avoided or stopped, simply by increasing the pressure of the liquid before it enters the suction nozzle of the pump. This will ensure that the pressure in the eye area does not fall below the vapor pressure, and therefore no vapor bubbles will be created and no cavitation will exist.

Much of the critical pressure drop that is created as the liquid moves into the eye of the impeller can be attributed simply to the loss of energy of a liquid moving from a static environment (the pump suction) to a dynamic environment in the rotating impeller. However, other design factors may occasionally play a part, such as the entrance angles of the impeller vanes as they relate to the velocity of the liquid.

4.4 Net positive suction head

The Pressure Energy needed to avoid the formation of vapor bubbles in the eye of the impeller in the cavitation process, is referred to as the Net Positive Suction Head (NPSH). The design criteria of each impeller require the supply of a minimum level of NPSH for its optimum performance, and are identified as the Net Positive Suction Head Required. It is strictly a function of the pump design and its rotational speed.

The pressure energy required by the pump is made available from the system in which the pump operates. In this form it is identified as the NPSH Available and is solely a function of the system design. Consequently, to avoid Cavitation damage, the NPSH Available must be greater than the NPSH Required.

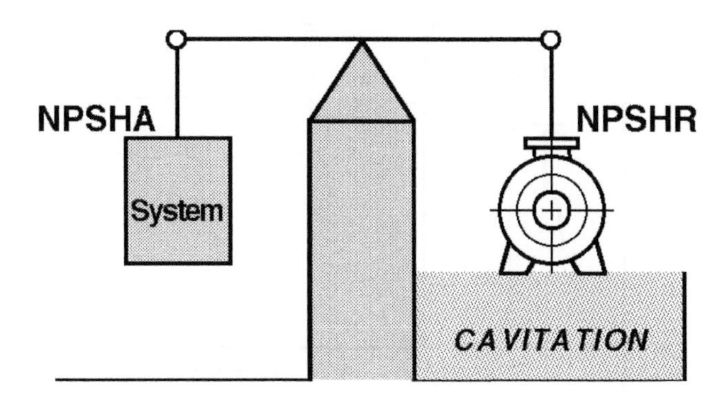

Figure 4.2: NPSH balance diagram

Therefore when Cavitation occurs in a pump, there are only two possible solutions:

- Decrease the NPSH Required, or
- Increase the NPSH Available.

In exploring possible cures for cavitation, it is interesting to note that (unless there has been a major selection or design flaw) most pumps cavitate because there is a pressure differential between the NPSHA and the NPSHR of less than a few feet. Consequently it is seldom necessary to make a major change to eliminate the problem.

4.4.1 NPSH Required by the pump

For well over 20 years every reputable Pump Manufacturer has conformed to a single testing standard to establish the NPSH required by a pump. That standard identifies the value of the Net Positive Suction Head required by the pump based on a 3% head drop. In other words, it is that amount of energy supplied to a pump that creates a reduction in the Total Head of no more than 3%. These factory tests are conducted at a constant flow rate in accordance with the Standards of the Hydraulic Institute and result in a curve similar to that shown in Figure 4.3.

A few specialty pumps in extremely critical applications are sometimes required to identify the NPSH required for a 1.0% head drop. On even more rare occasions the NPSH required at the 'Incipient Cavitation' point, is requested. This latter condition is essentially at that point where the first bubble can be heard imploding through special audio equipment. It should be stressed that these are not the standard 'off-

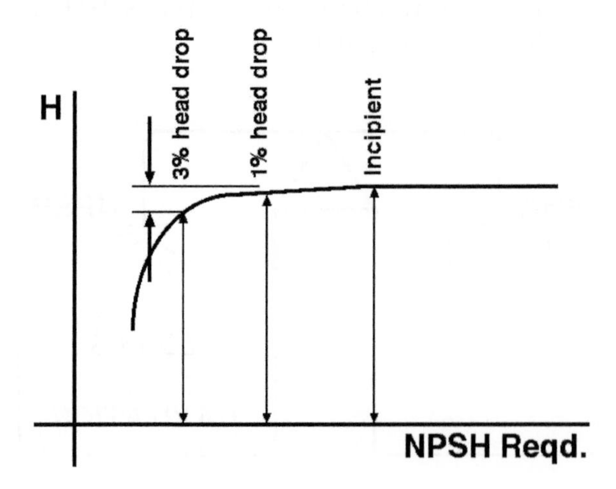

Figure 4.3: NPSH at constant flow rate

the-shelf' pump styles that are used by the vast majority of industry. These are special pumps only. All pump manufacturers design their standard pump range to operate with an NPSH value that is tested at a 3% head drop.

Consequently, every major pump manufacturer can identify the NPSH required by their pump when operating at a particular Head-Capacity condition. However it is important to recognize that, if no more than that amount is supplied, the pump will be cavitating, but at such a low level of energy that the resulting symptoms (i.e. noise, vibration and impeller damage) will be difficult to detect, and the long term detriment to the operation of the pump will be minimal.

There is a tendency in many areas to try and combat Cavitation by reducing the NPSH Required by the pump. It is worthwhile to realize that, to accomplish this, there are only a limited number of possibilities.

4.4.1.1 Increase the eye area of the impeller

As this option can cause more trouble that it solves by introducing recirculation difficulties, it is not recommended. It should only be considered as a last resort, and only with the full design involvement of the pump manufacturer.

4.4.1.2 Install a suction inducer

As very few pump manufacturers have suction inducers available, the practical application of this option will be severely limited. Even the few that are available must be approached with caution as they are likely to affect the pump performance at lower flows.

4.4.1.3 Use a double suction impeller

As the liquid flows into the impeller through two opposing eyes, a double suction impeller uses approximately 67% of the NPSH that is required by a single suction impeller in an equivalent size. This modification would necessitate a change of pump.

4.4.1.4 Use a slower speed pump

A slower speed requires less NPSH and will also necessitate a change to a much larger pump with a bigger impeller in order to accommodate the same performance conditions.

4.4.1.5 Use lower capacity pumps

A smaller, lower capacity pump also requires less NPSH, but will necessitate a change to multiple pumps in order to accommodate the same performance conditions.

4.4.1.6 Use a booster pump

Installed immediately upstream of the main pump, a booster pump must be able to operate at the same flow rate, but usually at a lower head, thus requiring less NPSH.

From this list of possibilities, you will note that there are specific concerns connected with the first two options, while the remaining ones require the installation of at least one new pump. Therefore to stop cavitation in most instances, the only really practical solution is to increase the NPSH available from the system.

4.4.2 NPSH available from the system

The NPSH Available from the System is relatively straightforward as it consists of only four absolute values.

$$NPSHA = Hs + Ha - Hvp - Hf$$

where: Hs is the Static Head over the impeller centerline,

Ha is the Head on the surface of the liquid in the suction tank,

Hvp equals the Vapor Pressure of the liquid, and

Hf is the Friction Losses in the Suction Line.

In the simple system shown, it can be seen that two factors will have a positive influence on the NPSH available, while two will have a negative influence. It is therefore apparent that, if a pump is cavitating, we should strive to increase the first two factors in the equation, and/or decrease the second two factors.

Once again, it should be stressed that a huge difference is not normally needed to eliminate cavitation. A few feet of NPSH will usually be enough. With this is mind, we can consider the four factors as they relate to a typical pump inlet system.

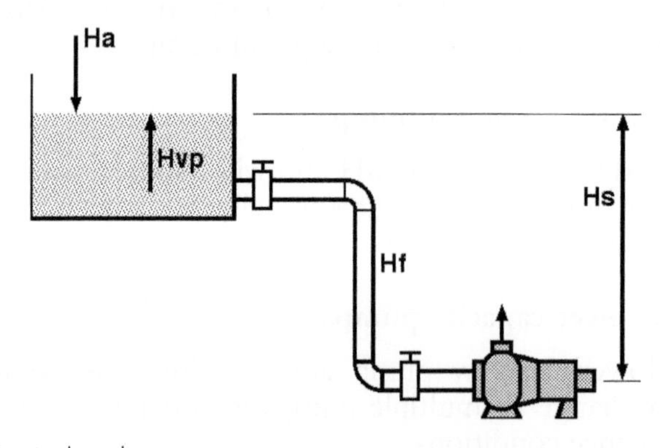

Figure 4.4: Suction tank and pump

4.4.2.1 Static head (Hs)

Increasing the static head available to the pump is a simple (?) matter of lowering the pump, or raising the suction tank, or the level in the tank. The physical movement of the tank or pump would usually be a costly proposition, yet the raising of the tank levels may be relatively cheap and simple, and can frequently cure the problem.

Of course, if the suction source happens to be an adjacent river or lake, there will be no control over the surface elevation and thus, the static head. The opposite problem of too much fluctuation is possible if the pump is being fed from a tidal source.

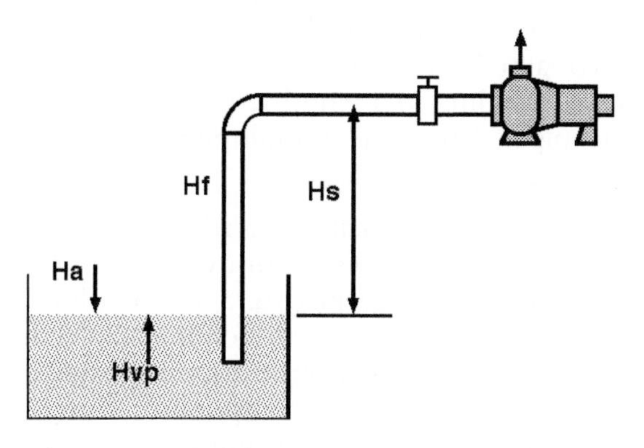

Figure 4.5: Below grade suction tank and pump

Where the pump is located above the level of the suction source, the Static Head will be a negative value, but all other considerations discussed above will remain the same.

4.4.2.2 Surface pressure (Ha)

Similarly the surface pressure can be a little tricky to change if the suction source is some body of water that resists control by mere mortals. It might be possible however to enclose a man-made tank and pressurize it, or even introduce a nitrogen blanket. Both of these possibilities are subject to the limitations of the particular service. For example, increasing the pressure inside a deaerator would defeat the whole function of that vessel and must therefore be considered impractical. However, as this pressure is one of only four factors in NPSHA formula, it is worthy of some consideration in certain installations.

4.4.2.3 Vapor pressure (Hvp)

The only way to reduce the vapor pressure of a liquid is to reduce its temperature. Under many operational conditions this will be unacceptable and can be ignored. Also, the extent of the temperature change needed to provide an appreciable difference in NPSHA will usually render this method ineffective.

4.4.2.4 Friction losses (Hf)

As pump inlet piping is notoriously bad in the vast majority of installations throughout the world, this is the area where significant improvements can often be realized. However, the tendency to shorten the length of suction piping simply to reduce friction losses should be resisted as it could deny the liquid the opportunity of a smooth flow path to the eye of the impeller. This, in turn, could cause turbulence and result in air entrainment difficulties that create the same symptoms as cavitation. To avoid this, the pump should be provided with a straight run of suction line in a length equivalent to 5 to 10 times the diameter of the pipe. The smaller multiplier should be used on the larger pipe diameters and vice versa.

The most effective way of reducing the friction losses on the suction side is to increase the size of the line. For example, friction losses can be reduced by more than 50% by replacing a 12 inch line with a 14 inch line. Exchanging a 6 inch line for an 8 inch line can reduce the friction losses by as much as 75%. Reduction in friction losses can be achieved even with the same line size by incorporating long sweep elbows, changing valve types and reducing their number.

Suction Strainers that are left over from the commissioning stage of a new plant can also be a problem. The blockage in the strainer basket steadily increases the friction loss to an unacceptable level. Even when a strainer is considered necessary in the process, it can frequently be located downstream of the pump and the pump selected to handle the solid sizes expected.

4.4.2.5 Sample NPSHA calculation

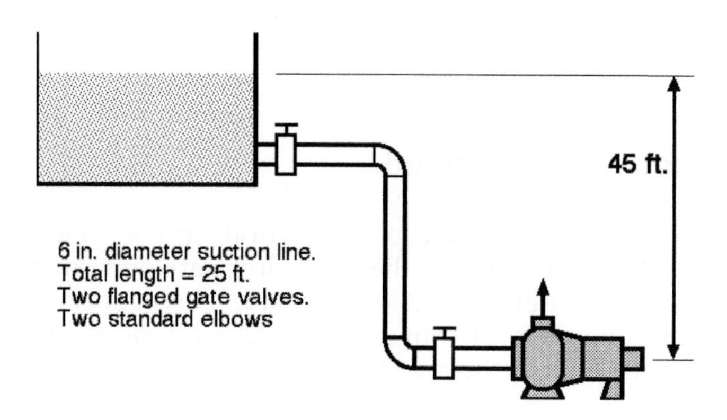

Figure 4.6: Diagram for NPSH calculation

Liquid to be pumped	Water at 140° F
Flowrate required through the pump	340 gpm
Calculated suction line size	6 inches

NPSHA = Hs + Ha − Hvp − Hf

Static Head (Hs)
The vertical distance from the elevation of the free surface of the liquid in the supply tank to the horizontal centerline of the impeller. = 45 ft.

Surface Pressure (Ha)
Absolute head on the surface of the liquid in the supply tank.
= Surface pressure x 2.31 / s.g.
= 14.7 psia x 2.31 / 0.985 = 34.474 ft.

Vapor Pressure (Hvp)
Vapor pressure expressed in terms of feet of head.
= 2.889 x 2.31 / 0.985 = 6.775 ft.

Friction Loss (Hs)
Piping Losses based on a flow rate of 340 gpm
(from Friction Loss Tables for 6 inch line) = 0.806 ft per 100 feet length
Losses in 25 ft = 0.806 x 25 / 100 = 0.2 feet

Valves and Fittings Losses based on a Flow Rate of 340 gpm
(from Resistance Coefficient Tables and Friction Loss Tables)

Velocity Head (V2/2g)	= 0.222	
K-Factor on Flanged Gate Valve	= 0.11	
K-Factor on Flanged Standard Elbow	= 0.29	
Friction Loss in 2 Valves	= 0.222 x 0.11 x 2	= 0.049 feet
Friction Loss in 2 Elbows	= 0.222 x 0.29 x 2	= 0.129 feet
	Total Friction Losses	= 0.378 feet

NPSHA	=	Hs	+	Ha	-	Hvp	-	Hf
	=	45	+	34.474	-	6.775	-	0.378
	=	**72.321 feet.**						

4.5 Suction specific speed

In some industries, the concept of Suction Specific Speed (Nss) has been introduced to compare the ideal flow rate and rotational speed with the NPSH required at that flow rate. This renders the NPSH a dimensionless number for convenient comparison of the hydrodynamic conditions that exist in the eye of the impeller.

$$Nss = \frac{RPM \times Q^{0.5}}{NPSHR^{0.75}}$$

where RPM = Pump rotational speed

Q = Flow at BEP in GPM

NPSHR = NPSH required at BEP in feet

The suction specific speed is calculated from the information on the manufacturer's pump performance curve and only at the Best Efficiency Point which is usually on the maximum diameter impeller. Consequently, a single line curve may not always be an appropriate reference, and a composite pump curve as shown in Figure 2.8 should be used. It is also further assumed that the Best Efficiency Point reflects the flow for which the eye of the impeller was originally designed.

When a double suction impeller is being considered, the flow (Q) in the above equation should be divided by two as the intent is to compare the performance in each individual impeller eye.

In many applications, the ability to use a pump with a low NPSH requirement would prove to be very beneficial in the physical design of the system. However, if this is carried to the extreme in pump design it has proved to cause recirculation problems within the impeller (see 4.6.1 below). This is particularly the case as it relates to operation of the pump at flows which may be much lower than the BEP. The use of suction specific speed provides a convenient method of identifying when such a condition may occur.

As the NPSH required is reduced, the value of the suction specific speed will increase. However, it has been noted that there is a tendency towards a decrease in pump reliability when the suction specific speed exceeds 11,000.

4.6 Confusing conditions

The reason that Cavitation continues to be a difficult problem to correct on a consistent basis, is that the classic symptoms of Cavitation are shared by three other conditions. This means that, when we experience the unique noise and high vibration levels, they could also be caused by Suction or Discharge Recirculation or by Air Entrainment, all of which have little to do with Cavitation or Suction Pressure.

4.6.1 Suction recirculation

This condition results from various types of instability such as turbulence, backflow circulation and swirling actions that can occur in the impeller when operating the pump at a low flow rate. Sometimes referred to as 'separation' or 'hydrodynamic' cavitation, these flow patterns tend to double back on themselves under low flows. Unfortunately, the flow rate at which this occurs will vary from one impeller to the next. Frequent occurrences at flows lower than 30% of the B.E.P. have been identified, while others have it tagged as high as 80%.

While the petrochemical industries favor a model that identifies recirculation taking place at the eye of the impeller, physical evidence in other industries shows the pitting damage almost halfway along the vane as shown in Figure 4.9. It would also appear as though the impeller design contributes to a condition where that damage could be on either the leading or the trailing edge of the vane.

In a nutshell, suction recirculation happens when the pump is operating at low flows, and the pitting damage normally takes place about halfway along the vanes.

4.6.2 Discharge recirculation

Discharge Recirculation is a very similar occurrence that results in pitting damage at the tip of the vanes and sometimes at the cut-water of the casing. It too can be caused by operating the pump at low flow rates. A similar type of damage can also be caused by recirculation between the tip of the impeller vanes and the cut-water of the casing when the radial clearance between these points is inappropriate.

4.6.3 Air entrainment

Air entrainment defines a variety of conditions where the vapor bubbles are already in the liquid before it reaches the pump. When they arrive in the eye of the impeller, exactly the same thing happens as if they were created at that point. In other words, the vapor is subjected to the increasing pressure at the start of the vanes and are then imploded, causing the identical damage as cavitation, and at the same location.

This condition can often be a result of pumping fermenting liquids or foaming agents found in a wide variety of industries. It can also be a result of pumping a liquid, such as condensate, that is close to its boiling point.

However, air entrainment is most frequently caused by turbulence in the suction line, or even at the suction source. For example, the kind of conditions identified in Figure 4.7, will cause turbulence in the suction tank that will entrain vapor bubbles into the line leading from that tank to the pump suction.

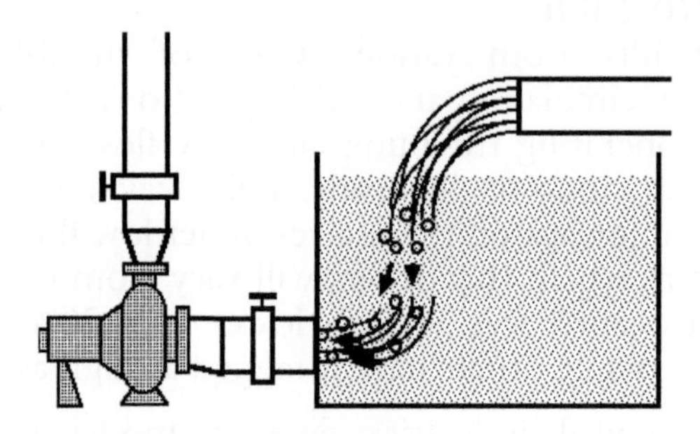

Figure 4.7: Effect of turbulence in suction tank

A similar condition can occur if the pump is drawing suction from a tank in which an agitator or fluid mixer is operating. These problems can frequently be minimized by the use of appropriate baffles in the tanks, if such a condition is feasible.

Turbulence in the suction lines to a pump can also be created by using too many elbows in the line. Even one elbow located directly onto the suction flange of the pump can create enough turbulence to cause air entrainment. If there are two elbows close to each other in the suction piping in different planes, the liquid will exit the second elbow in a swirling fashion that will cause considerable turbulence. This will create an air entrainment problem for the pump by causing pockets of low pressure in the liquid flow in which vaporization can occur.

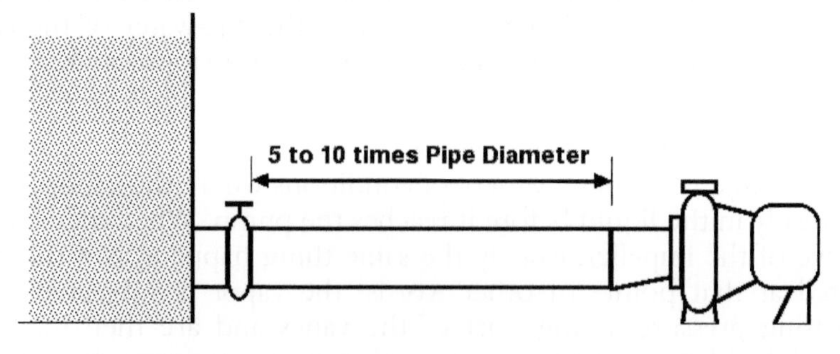

Figure 4.8: Suction pipeline

The ideal situation is to provide the suction side with a straight run of pipe, in a length equivalent to 5 to 10 times the diameter of that pipe, between the suction reducer and the first obstruction in the line. This will ensure the delivery of a uniform flow of liquid to the eye of the impeller and avoid any turbulence and air entrainment.

As air entrainment causes the same pitting damage to the impeller in precisely the same location as cavitation, it can be a little confusing, particularly as both can occur simultaneously in the same service. However, a quick comparison of the NPSHA and NPSHR, combined with a visual review of the piping characteristics will usually help identify the root cause of the so-called 'cavitation' and solve the air entrainment problem.

4.7 Similarities and differences

Cavitation, Air Entrainment and Recirculation all result in pitting damage on the impeller caused by the formation and subsequent collapse of vapor bubbles. The difference between them lies in the method by which the bubbles are formed and the location of their resultant implosions as shown in Figure 4.9.

As the severity of all these conditions increases, the noise, vibration and impeller damage will also increase. Under severe conditions, the pitting damage will spread throughout the impeller and may also extend to the casing.

All these conditions share some similar symptoms. As a consequence,

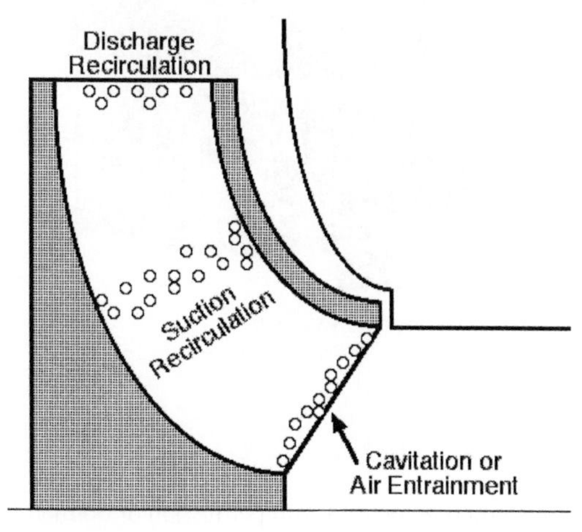

Figure 4.9: Bubble implosion locations

they can be diagnosed incorrectly. However, they are caused by three separate conditions and, by focusing on these root causes, an accurate diagnosis can be simplified.

It must be recognized that the harmful effects on the impeller is only one consequence of these conditions. The bigger problems come from the subsequent vibration and its detrimental effects on seals and bearings.

4.8 Priming

Another important suction condition exists when a pump is operating on a suction lift. When the pump stops, there is a tendency for the liquid to run out of the suction pipe. If this occurs and the pump has to restart under this condition, it must be able to handle the air pocket that's now in the suction line.

The most popular method of dealing with this eventuality is with the use of a self-priming pump which is capable of freeing itself of entrained gas and resuming normal pumping without any attention.

These pumps have a suction reservoir cast integrally with the pump casing to retain a certain volume of liquid even when the suction line is

Figure 4.10: Self-priming pump (Reproduced with permission of Gorman-Rupp Pump Company)

drained by gravity. When the pump restarts, it recirculates that same liquid through the priming chamber until all the air has been passed through and normal pumping is reestablished.

4.8.1 Self-priming pump layout

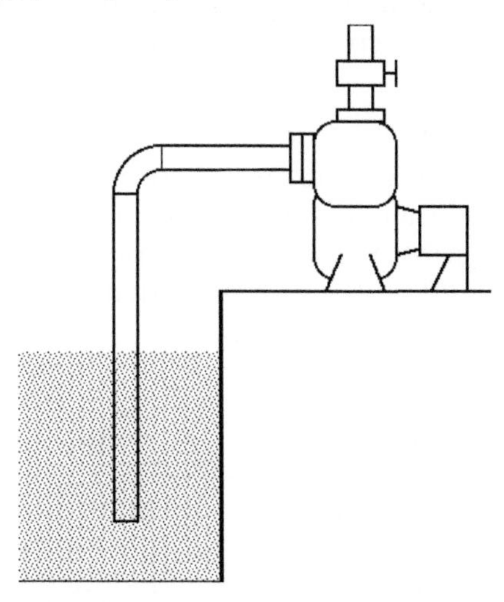

Figure 4.11: Self-priming pump system diagram

This standard arrangement will usually locate a foot valve at the bottom end of the suction line to prevent the liquid draining back into the sump. Unfortunately, the simple design of these foot valves renders them susceptible to sticking in the open position and allowing the pipe to empty.

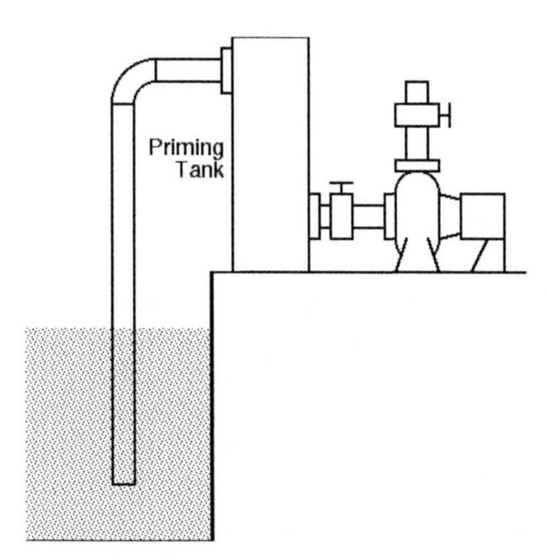

Figure 4.12: Priming tank pump system

4.8.2 Centrifugal pump with priming tank

This priming tank acts in a manner similar to that of the suction trap in a self-priming pump, and must be sized so that it contains 3 times the volume of the suction line. When the pump starts to empty the tank, it creates enough of a vacuum in the priming tank to draw the suction line full again before the tank empties. During this time it will also be capable of supplying sufficient NPSH to the pump.

4.8.3 Air ejector system

An air ejector system can be automated to use available compressed air to vacate the entrained air in the suction line and pump prior to the pump startup. By creating a vacuum in the pump, it will draw the liquid into the suction line and fill the pump. At that point the pump will start in a fully primed condition.

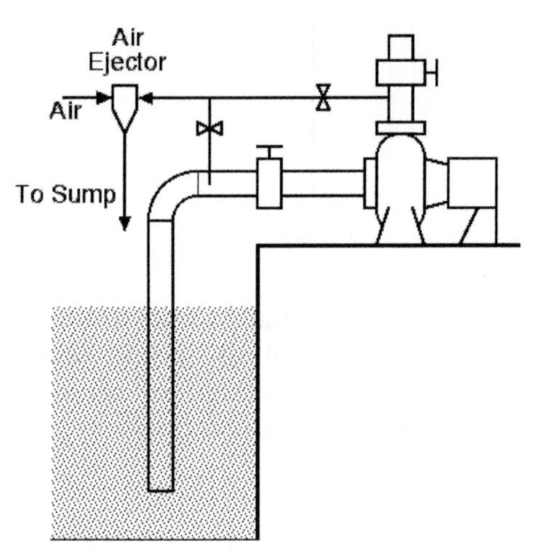

Figure 4.13: Air ejector priming system

4.9 Submergence

Submergence is the static elevation difference between the free surface of the liquid and the centerline of the impeller in a vertical shaft pump. Inadequate submergence causes random vortices that permit air to be drawn into the pump, causing increased vibration and reduced life. This required submergence is completely independent of the NPSH required by the pump.

Inadequate sump design frequently causes serious pump problems but many sources, such as the Hydraulic Institute Standards, provide guidelines to pump arrangements and clearances between pumps, floor and walls.

The fundamental requirement of a good sump design is that the liquid must not encounter sharp turns or obstruction which may generate a vortex as it flows to the pump suction.

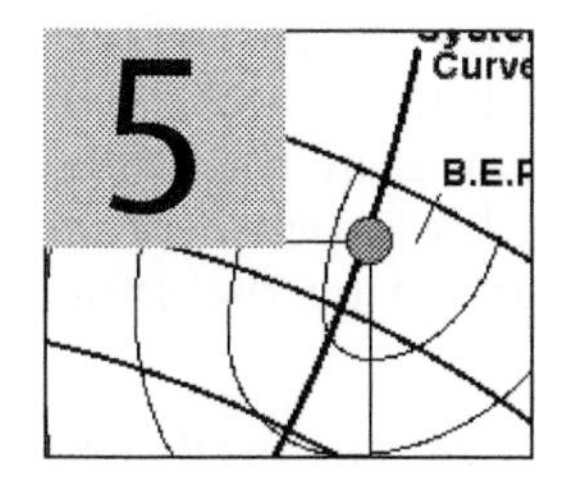

Pump selection and purchasing

5.1 Pump selection factors

One of the earliest decisions in the traditional design of a system is the selection of the Head and Capacity for which the pump should be sized. As was discussed in Chapter 3, this usually results in the selection of the pipe sizes to be used which, in turn, determines the friction losses to be overcome. With the introduction of various computer software programs, the selection of pump performance and system performance is now being combined. By using these programs to effectively compare piping costs (and all related fittings), pump costs and power costs for various pipe sizes, the optimum combination can be determined.

Regardless of how the Head – Capacity conditions may be selected, much more information is required to ensure an optimal selection of pump style. For example, let's consider the liquid itself.

- Is it corrosive?
- Is it abrasive?
- Are there solid particles and, if so, what size and percentage are they?
- Is it a viscous liquid and if so, what is the viscosity?
- Does it tend to crystallize or otherwise solidify?
- What is the vapor pressure?
- Is it temperature sensitive?

If the liquid to be pumped is cold, clean potable water, most people are sufficiently aware of the character of that liquid to understand that none of the above factors will play a significant part in the pump selection process.

However, even 'water' comes in different forms, from Condensate to Brine, which could require a wide variety of corrosion resistant materials. Even Sea Water can vary in corrosiveness from one part of the ocean to another. In addition, the abrasiveness of the water in a particular mine may dictate the use of a rubber lining in the Mine Dewatering pump purchased, while other mines have less expensive cast iron pumps performing what, at first glance, appears to be the same service.

It is also worthwhile to remember that numerous new chemicals are now being introduced to many industrial processes, therefore a detailed knowledge of the liquid should never be assumed.

Consequently, the following items should be considered the minimum data required for the selection of an appropriate centrifugal process pump to suit the service for which it is intended.

1. The liquid to be pumped.
2. Flow rate required.
3. Total Dynamic Head.
4. Net Positive Suction Head available.
5. Operating temperature.
6. Specific Gravity.
7. Nature of the liquid. (See listing above.)
8. Operational experience.

5.1.1 Operational experience

The last item on the above list, 'Operational Experience', is the kind of information that is rarely found recorded anywhere and usually needs to be uncovered by exploring the following topics.

- Is it on a nonstop duty?
- What is the start/stop frequency?
- Will it be used on one operation only?
- Will it pump a variety of liquids?
- What kind of controls are on the discharge?
- Will the flow be restricted at any time?
- Is it on an open or closed loop or a transfer system?

If it is a replacement pump, other questions should include;

- What pump model was used?

- At what speed was it running?
- What materials of construction were used?
- What type of mechanical seal was used?
- What types of auxiliary systems were used?
- What kind of operational/maintenance record was experienced?

Once this type of information has been established the selection of the pump can proceed with the type of unit required and then the best hydraulic fit for the system.

5.1.2 End user data sheets

Many industries provide their suppliers with standardized data sheets which have evolved over the years. While a well designed data sheet will provide much of the information reviewed above, they generally do not supply any system information or previous operational experience, both of which are vital for an optimum pump selection. It is strongly recommended that the system designer and equipment supplier cooperate as much as possible to ensure that the best possible selection ensues.

One of the most widely used data sheets is the Centrifugal Pump Data Sheet incorporated in the API Standard 610 for Centrifugal Pumps for General Refinery Service that is published by the American Petroleum Institute. This is also available in SI units, and both sheets are provided here for the benefit of the reader. Many companies have modified this data sheet to suit their own specific requirements.

5.2 System operating considerations

An essential part of the 'Operational Experience' is a detailed knowledge of the system in which the pump will operate.

5.2.1 Closed loop systems

When a pump is required to perform in a closed loop where the pump discharges through the system and back to it's own suction nozzle, the static head is effectively zero and the system characteristics will remain fairly constant. Consequently, the system curve will experience little movement, and the pump selection needs only to generate a pump performance curve that intersects with the system curve at the required Design Point. With such a stable operation, there will be little movement of either the pump curve or the system curve, and the operating point will remain at the Design Flow throughout the life of

ISO 13709 (API 610 9TH)
CENTRIFUGAL PUMP
PROCESS DATA SHEET
ISO STANDARDS (4.2)
S.I. UNITS (4.3)

PAGE 1 OF
JOB NO. ITEM NO.(S)
REQ / SPEC NO. /
PURCH ORDER NO. DATE
INQUIRY NO BY

1 APPLICABLE TO: O PROPOSALS O PURCHASE [O] AS BUILT
2 FOR UNIT
3 SITE SERVICE
5 NOTES: INFORMATION BELOW TO BE COMPLETED: O BY PURCHASER ☐ BY MANUFACTURER [O] BY MANUFACTURER OR PURCHASER
6 O DATA SHEETS (6.1.1) REVISIONS

7	ITEM NO.	ATTACHED	ITEM NO.	ATTACHED	ITEM NO.	ATTACHED	NO.	DATE	BY
8 PUMP		O		O		O	1		
9 MOTOR		O		O		O	2		
10 GEAR		O		O		O	3		
11 TURBINE		O		O		O	4		
12 APPLICABLE OVERLAY STANDARD(S):							5		

13 O OPERATING CONDITIONS (5.1.3)
14 CAPACITY, NORMAL (m³/h) RATED (m3/h)
15 OTHER
17 SUCTION PRESSURE MAX./RATED / (bar)
18 DISCHARGE PRESSURE (bar)
19 DIFFERENTIAL PRESSURE (bar)
20 DIFF. HEAD (m) NPSHA (m)
21 PROCESS VARIATIONS (5.1.4)
22 STARTING CONDITIONS (5.1.4)
23 SERVICE: O CONT. O INTERMITTENT (STARTS/DAY)
24 O PARALLEL OPERATION REQ'D (5.1.13)
25 O SITE DATA (5.1.3)
27 LOCATION: (5.1.30)
28 O INDOOR O HEATED O OUTDOOR O UNHEATED
29 O ELECTRICAL AREA CLASSIFICATION (5.1.24 / 6.1.4)
30 CL GR DIV
31 O WINTERIZATION REQ'D O TROPICALIZATION REQ'D.
32 SITE DATA (5.1.30)
33 O ALTITUDE (m) BAROMETER (bar)
34 O RANGE OF AMBIENT TEMPS: MIN/MAX. / (°C)
35 O RELATIVE HUMIDITY: MIN / MAX / (%)
36 UNUSUAL CONDITIONS: (5.1.30) O DUST O FUMES
37 O OTHER
38
39
40 O DRIVER TYPE
41 O INDUCTION MOTOR O STEAM TURBINE O GEAR
42 O OTHER
43
44 O MOTOR DRIVER (6.1.1 / 6.1.4)
45 [O] MANUFACTURER
46 ☐ (KW) ☐ (RPM)
47 ☐ FRAME [O] ENCLOSURE
48 [O] HORIZONTAL [O] VERTICAL [O] SERVICE FACTOR
49 [O] VOLTS/PHASE/HERTZ / /
50 O TYPE
51 O MINIMUM STARTING VOLTAGE (6.1.5)
52 [O] INSULATION O TEMP. RISE
53 [O] FULL LOAD AMPS
54 [O] LOCKED ROTOR AMPS
55 [O] STARTING METHOD
56 [O] LUBE
58 BEARINGS (TYPE/NUMBER):
59 ☐ RADIAL /
60 ☐ THRUST /
61 ☐ VERTICAL THRUST CAPACITY
62 UP (N) DOWN (N)
63
64
65

O LIQUID (5.1.3)
LIQUID TYPE OR NAME
O HAZARDOUS O FLAMMABLE O (5.1.5)

	MIN.	NORMAL	MAX.
PUMPING TEMP (°C)			
VAPOR PRESS. (bar)			
RELATIVE DENSITY (SG):			
VISCOSITY (PaS)			

SPECIFIC HEAT, Cp (kJ/kg °C)
O CHLORIDE CONCENTRATION (PPM)
O H_2S CONCENTRATION (6.5.2.4) (PPM) WET (5.2.1.12c)
CORROSIVE / EROSIVE AGENT (5.12.1.9)

MATERIALS
O ANNEX H CLASS (5.12.1.1)
O MIN DESIGN METAL TEMP (5.12.4.1) (°C)
O REDUCED HARDNESS MATERIALS REQ'D. (5.12.1.11)
☐ BARREL/CASE IMPELLER
☐ CASE/IMPELLER WEAR RINGS
☐ SHAFT
☐ DIFFUSERS

[O] PERFORMANCE:
PROPOSAL CURVE NO. ☐ RPM
☐ IMPELLER DIA. RATED MAX. MIN. (mm)
☐ IMPELLER TYPE
☐ RATED POWER (kW) EFFICIENCY (%)
☐ MINIMUM CONTINUOUS FLOW:
THERMAL (m³/h) STABLE (m³/h)
☐ PREFERRED OPER. REGION TO (m³/h)
☐ ALLOWABLE OPER. REGION TO (m³/h)
☐ MAX HEAD @ RATED IMPELLER (m)
☐ MAX POWER @ RATED IMPELLER (kW)
☐ NPSHR AT RATED CAPACITY (m) (5.1.10)
[O] SUCTION SPECIFIC SPEED
MAX/ACTUAL / (5.1.11)
[O] MAX. SOUND PRESS. LEVEL REQ'D (dBA) (5.1.16)
[O] EST MAX SOUND PRESS. LEVEL (dBA) (5.1.16)

O UTILITY CONDITIONS

ELECTRICITY	VOLTAGE	PHASE	HERTZ
DRIVERS			
HEATING			

SYSTEM VOLTAGE DIP O 80% O OTHER (6.1.5)

STEAM	MAX. PRESS.	MAX. TEMP.	MIN. PRESS.	MIN. TEMP.
DRIVERS				
HEATING				

COOLING WATER: (5.1.19) SOURCE
SUPPLY TEMP. (°C) MAX RETURN TEMP. (°C)
NORM. PRESS. (bar) DESIGN PRESS. (bar)
MIN. RET. PRESS. (bar) MAX. ALLOW. D.P. (bar)
CHLORIDE CONCENTRATION: (PPM)

05/01 SHT 1 610 ISO.XLS REV 0 PROCESS DATA SHEET

Figure 5.1: API data sheet in U.S. units (Reproduced with permission of The American Petroleum Institute)

PAGE 1 OF

ISO 13709 (API 610 9TH)
CENTRIFUGAL PUMP
PROCESS DATA SHEET
U.S. STANDARDS(4.2)
U.S. UNITS (4.3)

JOB NO. ITEM NO.(S)
REQ / SPEC NO. /
PURCH ORDER NO. DATE
INQUIRY NO BY

1 APPLICABLE TO: O PROPOSALS O PURCHASE O AS BUILT
2 FOR UNIT
3 SITE SERVICE
5 NOTES: INFORMATION BELOW TO BE COMPLETED: O BY PURCHASER ☐ BY MANUFACTURER O BY MANUFACTURER OR PURCHASER

6 O DATA SHEETS (6.1.1) — REVISIONS

7	ITEM NO.	ATTACHED	ITEM NO.	ATTACHED	ITEM NO.	ATTACHED	NO.	DATE	BY
8 PUMP		O		O		O	1		
9 MOTOR		O		O		O	2		
10 GEAR		O		O		O	3		
11 TURBINE		O		O		O	4		
12 APPLICABLE OVERLAY STANDARD(S):							5		

13 O OPERATING CONDITIONS (5.1.3)
14 CAPACITY, NORMAL (GPM) RATED (GPM)
15 OTHER
17 SUCTION PRESSURE MAX./RATED / (PSIG)
18 DISCHARGE PRESSURE (PSIG)
19 DIFFERENTIAL PRESSURE (PSI)
20 DIFF. HEAD (FT) NPSHA (FT)
21 PROCESS VARIATIONS (5.1.4)
22 STARTING CONDITIONS (5.1.4)
23 SERVICE: O CONT. O INTERMITTENT (STARTS/DAY)
24 O PARALLEL OPERATION REQ'D (5.1.13)

25 O SITE DATA (5.1.3)
27 LOCATION: (5.1.30)
28 O INDOOR O HEATED O OUTDOOR O UNHEATED
29 O ELECTRICAL AREA CLASSIFICATION (5.1.24 / 6.1.4)
30 CL GR DIV
31 O WINTERIZATION REQ'D O TROPICALIZATION REQ'D.
32 SITE DATA (5.1.30)
33 O ALTITUDE (FT) BAROMETER (PSIA)
34 O RANGE OF AMBIENT TEMPS: MIN/MAX / (°F)
35 O RELATIVE HUMIDITY: MIN / MAX / (%)
36 UNUSUAL CONDITIONS: (5.1.30) O DUST O FUMES
37 O OTHER
38
39

40 O DRIVER TYPE
41 O INDUCTION MOTOR O STEAM TURBINE O GEAR
42 O OTHER
43

44 O MOTOR DRIVER (6.1.1 / 6.1.4)
45 O MANUFACTURER
46 ☐ (HP) ☐ (RPM)
47 ☐ FRAME O ENCLOSURE
48 O HORIZONTAL O VERTICAL O SERVICE FACTOR
49 O VOLTS/PHASE/HERTZ / /
50 O TYPE
51 O MINIMUM STARTING VOLTAGE (6.1.5)
52 O INSULATION O TEMP RISE
53 O FULL LOAD AMPS
54 O LOCKED ROTOR AMPS
55 O STARTING METHOD
56 O LUBE
58 BEARINGS (TYPE/NUMBER):
59 ☐ RADIAL /
60 ☐ THRUST /
61 ☐ VERTICAL THRUST CAPACITY
62 UP (LBS) DOWN (LBS)
63
64
65

O LIQUID (5.1.3)
LIQUID TYPE OR NAME
O HAZARDOUS O FLAMMABLE O (5.1.5)

	MIN.	NORMAL	MAX.
PUMPING TEMP (°F)			
VAPOR PRESS. (PSIA)			
RELATIVE DENSITY (SG):			
VISCOSITY (cP)			

SPECIFIC HEAT, Cp (BTU/LB °F)
O CHLORIDE CONCENTRATION (PPM)
O H_2S CONCENTRATION (6.5.2.4) (PPM) WET (5.2.1.12c)
CORROSIVE / EROSIVE AGENT (5.12.1.9)

MATERIALS
O ANNEX H CLASS (5.12.1.1)
O MIN DESIGN METAL TEMP (5.12.4.1) (°F)
O REDUCED HARDNESS MATERIALS REQ'D. (5.12.1.11)
☐ BARREL/CASE IMPELLER
☐ CASE/IMPELLER WEAR RINGS
☐ SHAFT
☐ DIFFUSERS

O PERFORMANCE:
PROPOSAL CURVE NO. ☐ RPM
☐ IMPELLER DIA. RATED MAX. MIN (IN.)
☐ IMPELLER TYPE
☐ RATED POWER (BHP) EFFICIENCY (%)
☐ MINIMUM CONTINUOUS FLOW:
THERMAL (GPM) STABLE (GPM)
☐ PREFERRED OPER. REGION TO (GPM)
☐ ALLOWABLE OPER. REGION TO (GPM)
☐ MAX HEAD @ RATED IMPELLER (FT)
☐ MAX POWER @ RATED IMPELLER (BHP)
☐ NPSHR AT RATED CAPACITY (FT) (5.1.10)
O SUCTION SPECIFIC SPEED
MAX/ACTUAL / (5.1.11)
O MAX. SOUND PRESS. LEVEL REQ'D (dBA) (5.1.16)
O EST MAX SOUND PRESS. LEVEL (dBA) (5.1.16)

O UTILITY CONDITIONS

ELECTRICITY	VOLTAGE	PHASE	HERTZ
DRIVERS			
HEATING			

SYSTEM VOLTAGE DIP O 80% O OTHER (5.1.5)

STEAM	MAX. PRESS.	MAX. TEMP.	MIN. PRESS.	MIN. TEMP.
DRIVERS				
HEATING				

COOLING WATER: (5.1.19) SOURCE
SUPPLY TEMP. (°F) MAX. RETURN TEMP. (°F)
NORM. PRESS. (PSIG) DESIGN PRESS. (PSIG)
MIN. RET. PRESS. (PSIG) MAX. ALLOW. D.P. (PSI)
CHLORIDE CONCENTRATION: (PPM)

05/01 SHT 1 610USDS1.XLS REV 0 PROCESS DATA SHEET

Figure 5.2: API data sheet in S.I. units (Reproduced with permission of The American Petroleum Institute)

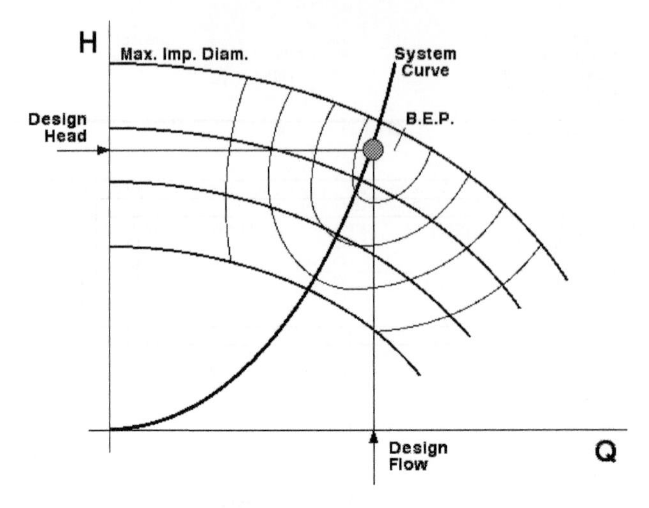

Figure 5.3: Closed loop performance curve

the pump. The only changes over time will be the result of wear on the pump and system.

In selecting the pump for such a service, the pump supplier will select the required pump type and then compare the hydraulic requirements of Head and Flow to the Characteristic Pump Performance Curves of various pumps of that type. The best selection will usually be considered to be that pump whose curve meets the system requirements as closely as possible to the Best Efficiency Point. The best selection will also usually have the Design Flow a little to the left of the Best Efficiency Point on the curve. This provides for a pump that will operate as efficiently and smoothly as possible under the conditions identified by the System Designer.

5.2.2 Batch transfer system

In a batch transfer type of system such as shown in Figure 5.4, the Total Dynamic Head is usually a constantly changing condition where the most dramatic change is that which is created by emptying the supply tank. This brings about a reduction in the liquid level in that tank and the equivalent increase in the Total Static Head that the pump must overcome.

In considering the three conditions identified in Figure 5.4, the pump will usually start with the Lowest Head when the suction source is full and the discharge tank is empty. When the pump has emptied the suction source, it achieves the Ultimate Head at the maximum Total Static Head, at which point it should be shut down.

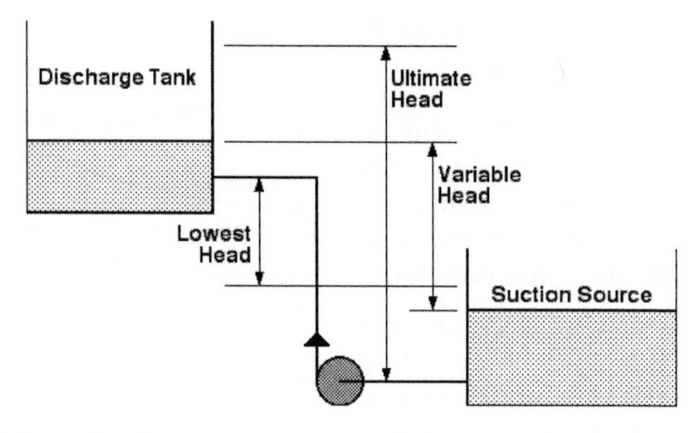

Figure 5.4: Batch transfer diagram

As soon as the pump starts, the level in the suction source begins to drop, thus increasing the Total Static Head through which the pumpage must be raised. This is referred to as the Variable Head and it has infinite possibilities for all points between the Lowest curve and the Ultimate curve as shown in Figure 5.5.

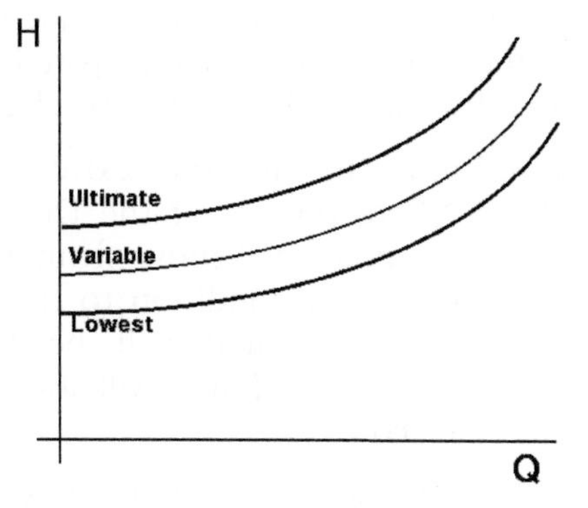

Figure 5.5: Batch transfer system curves

Unfortunately, many Engineers fall into the trap of automatically focusing on the Ultimate Head as it includes the largest value of Total Static Head, and is further considered as the 'worst' operating condition. The traditional thought process assumes that, if the pump can handle the 'worst' condition, it must therefore be able to handle all other conditions.

The reality is that, if the 'worst' condition is used as the pump selection point, it automatically becomes the 'best' condition, because that's the point at which the pump and system have been matched.

With this thought process, the 'worst' condition now becomes the system's Design Flow. Therefore, if we select Pump 'A' to operate at (or as close as possible to) it's Best Efficiency Point at the Design Flow, the relationship of the pump with all three systems will be as shown in Figure 5.6.

When the system starts on the Lowest System, the pump selected (Pump 'A') will now be operating well to the right of the Best Efficiency Point under very unstable conditions and at a much higher flow rate than was originally planned. From this point the Variable System will slowly move vertically upwards to the Ultimate System. This will bring the operating point back along the pump curve to the Design Flow and a smooth operating condition.

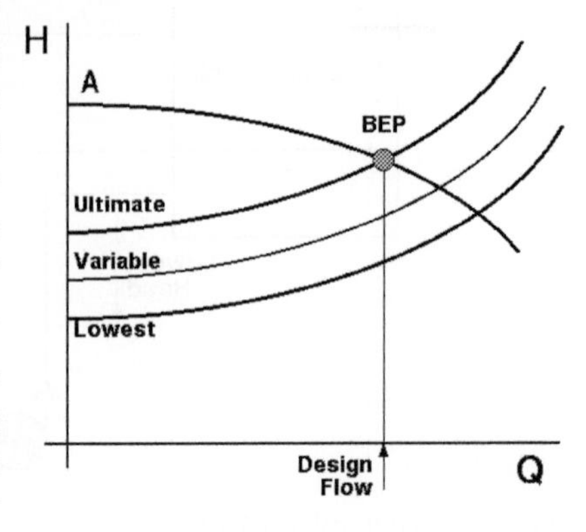

Figure 5.6: System curve with pump A

As a result of this decision, the pump will therefore be operating for most of the time under adverse conditions which will inevitably result in poor reliability and premature failure. In addition, the batch will be transferred at a much higher speed than was anticipated as the pump will always be operating at a flow rate higher than was selected.

If we move to the other extreme as shown in Figure 5.7, and select Pump 'B' to operate at the Design Flow on the Lowest System, the operating point will move from a reliable condition to a less efficient and less reliable condition to the left of the Best Efficiency Point. In this case, the batch will be transferred much slower than was anticipated, as the pump will always be operating a lower flow rate than was selected.

Under the worst of these cases, the pump selected for the Lowest System, may have a performance curve that is so flat, it does not intersect the Ultimate System at any point as shown in Figure 5.8. This

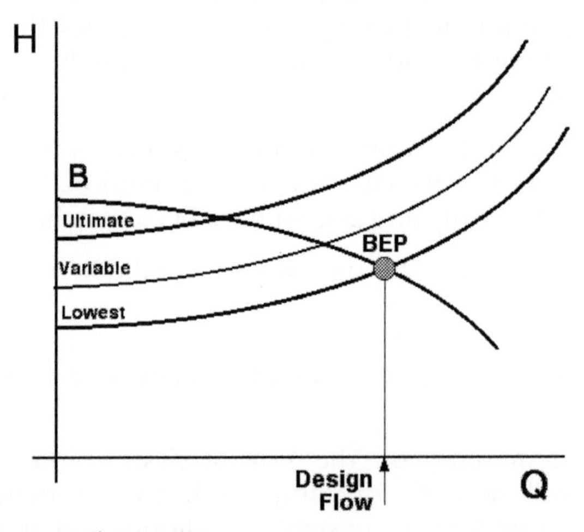

Figure 5.7: System curve with pump B

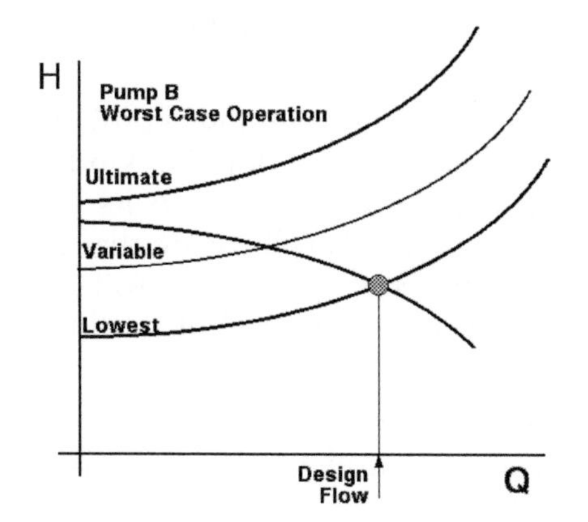

Figure 5.8: Pump B – worst case operation

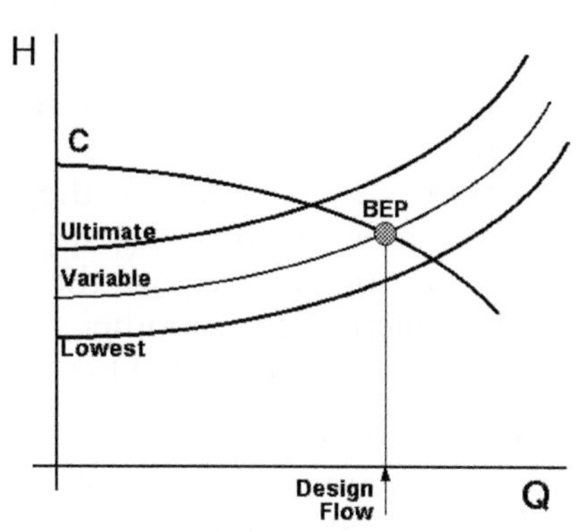

Figure 5.9: System curve with pump C

means that the Ultimate Head may never be reached, and the suction source never emptied.

However, if Pump 'C' was selected with the Design Flow on the Variable System, midway between the Lowest System and the Ultimate System, the optimum condition could be achieved. As shown in Figure 5.9, the operating point of Pump 'C' will start to the right of BEP and gradually move to the left of BEP. This will usually provide the most efficient and reliable option.

When selecting a pump for this type of system, it is essential to ensure that it can handle all the operating conditions to which it will be subjected. The only way to do that is to draw a system curve for all the different conditions and compare the pumps with all of these system possibilities.

5.2.3 Multiple destination systems

A similar situation is found in the type of application where one pump supplies a number of destinations at various locations and distances from the pump. Under such conditions, the system curve must be drawn for each service, and superimposed on the single pump curve as shown in Figure 5.10.

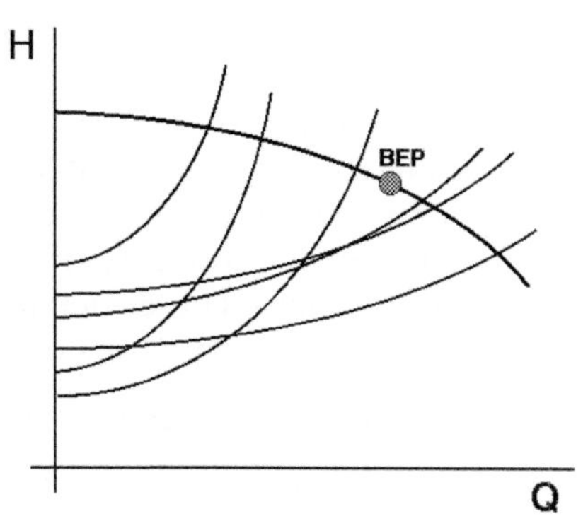

Figure 5.10: Multiple destination systems

With varying destinations for the liquid, it can be assumed that there will be different static heads involved and, with the destinations being at different distances, there will also be different levels of friction losses.

The selection of a single pump for such a system will impose a different flow rate for every system. The best pump selection for this kind of operation is the one where the pump will operate closest to its best efficiency point for the longest period of time. This will ensure that the pump will not only require the lowest horsepower draw, but will also operate quieter and smoother and with an increased level of reliability throughout it's operational life.

Depending on the shape of the various system curves and the demands of all the systems, it may be beneficial to consider using a Variable Speed Drive on a pump in this kind of service.

5.3 Price evaluation

When we buy value instead of price for our own consumption, we only have to justify the buying decision to ourselves, and that's usually an easy thing to do. The justification of a higher priced pump to a colleague is not always as straight forward. This is particularly so when the other's function is perceived as being expenditure reduction. Such a focus is likely to eliminate all other factors from consideration. The difficulty is compounded if the colleague has limited knowledge of pumps and the systems in which they operate. The easy way out is to succumb to the simplistic justification that, if all the pumps meet the specs, and the suppliers guarantee they will do the job, then they are all acceptable and the cheapest pump is the best buy.

There are three problems with this assumption.

1. **The pump may not fully meet the requirements of the specification.**

   Having witnessed the situation from both sides, I have rarely seen a statement from a supplier to the effect that the equipment being offered will fully meet the specification requirements. Instead, we usually find a listing of comments pertaining to various paragraphs within the specification. Frequently, many of the details are left to the integrity of the relationship between the purchaser and supplier. These details can often become the subject of intense and acrimonious negotiations should a problem ever arise.

2. **The specification is not appropriate.**

   This is a more serious situation and happens frequently when incomplete information is provided. This does not refer to any vindictive withholding of data, but rather to the limitations of knowledge of the field conditions and the understanding of how these conditions might impact the pump performance and reliability. This underscores the need for all the questioning identified in the early stages of this Chapter.

It should also be noted that, just because a centrifugal pump has been purchased to deliver 500 gallons per minute, it will not necessarily do that in the field, as it is under constant control of the system.

All pump problems arise from either internal or external stresses. The internal stresses mostly occur as a result of upset hydraulic conditions which are rarely discussed in any specification, regardless of how integral they are to the system in which the pump must operate. The external stresses come from inappropriate installation or operation, and they too are rarely acknowledged, particularly in the specification.

3. **Suppliers do not guarantee field performance.**

   This is the most important point. The standard pump manufacturer's guarantee covers defective material and workmanship of parts of its own manufacture. They will also agree to meet (within certain tolerances and under factory test conditions) all specified conditions of Head, Capacity, NPSH and sometimes Vibration, with the stated levels of Efficiency and Power Draw offered.

   In the event that anything goes wrong in the field, the first reaction is to take the pump back to the factory and test it under controlled conditions. This is quite understandable and justified, as the supplier has no control over the installation and operating conditions in the field. Consequently, field performance is rarely, if ever, guaranteed.

5.3.1 Traditional pricing negotiation

Otherwise known as 'Your price is too high!'. This is the standard response to any equipment salesperson from any customer anywhere in the world, and while the language might change with the product and the country, the idea is exactly the same. Many salespeople have not yet realized that this statement is simply a set piece in the formal script of the negotiation process which has become part of the pump purchasing scenario over the years. Any Purchasing Officer worth their salt will use that phrase regardless of the specific numbers in front of them.

The sad part is that, over the years, it has created a trend towards price cutting that has served neither manufacturer nor user. Witness the number of companies that no longer serve certain markets where this practice is still in wide usage; resulting in some good quality products that are no longer available in these markets. This in turn, leaves those markets with inappropriate and usually inefficient equipment with which to transfer and process the liquids needed in their operation.

The long term consequence of this scenario usually finds these same end users negotiating low bids on the prices of the over abundance of spare parts they need to keep their pumps operational. Many of which are now supplied by third party organizations who, in turn, can take no responsibility for any changes in hydraulic operation of the pumps for which they provide the parts.

5.3.2 Value based purchasing

Value Based Purchasing helps us to ensure that the pumps we buy are the best available in the market and the ones most suited to the operation for which they are being purchased.

It should be noted that the detailed evaluation and comparison of pump quotes cannot be considered an exact science as it will always be necessary to make a subjective evaluation about the accuracy of the data presented. In this area, some previous experience with pumps and potential suppliers will be invaluable. However, the overall consideration must continue to be the best long term value for the money.

To do that, we must consider specific aspects of the equipment being purchased.

1. Hydraulic suitability to the service.
2. Efficiency of operation.
3. Mechanical suitability to the service.

Most companies already include some kind of evaluation for the first two factors in their considerations. The only comment that can be made would be in relation to the hydraulic suitability of the pump for the service. This can only be established if consideration is given to every variation of the system curve as discussed earlier in this Chapter.

With respect to Efficiency of operation, an evaluation of power cost is fairly standard, and is based on the particular value of power which is considered appropriate in each plant. To ensure a real appreciation of the value of an efficient pump, it is strongly recommended that the actual power consumption cost for each pump under consideration is calculated. Do not be tempted into only calculating the difference in efficiency quoted, as it tends to provide a false impression.

When considering each of the various factors involved in the mechanical suitability of the pump for the service, the concept is to give a higher evaluation to the factors that will encourage or even ensure increased reliability of the pump.

The most straightforward method is to assess a penalty to any factors in the evaluation which can be considered less than ideal. The penalty can take a variety of forms and may often be fairly subjective. However, the more objective the assessment, the more reliable the evaluation will be.

5.3.3 Factors involved in mechanical suitability

5.3.3.1 Mechanical seals

The most obvious example is the mechanical seal as it is that part of a pump that most frequently fails. Consequently it is highly recommended that a decision on which seal supplier, model and materials is needed should be finalized up front and included in the pump specification. This helps all pump vendors and ensures you get what you really need, rather than the cheapest available on the market. However, if a variety of seals have been offered in the various bids received, there are a number of design features which can be clearly identified and evaluated.

(a) For example, the cartridge design is simpler and safer to use than the component seal. It protects the seal faces against damage and reduces the time and skill level required for installation. It can be argued that a sound evaluation assessment for not providing a cartridge seal may be the addition of the cost of an additional spare seal of the type originally quoted.

(b) Any seal which has a dynamic secondary seal will inevitably cause fretting corrosion which leaves a damaged shaft. With this type of a seal, the shaft must be protected by a sleeve which will require frequent replacement. A non-fretting design will not need a shaft sleeve, thus eliminating one spare part. It also allows a larger diameter (and thus a stronger) shaft.

One suggestion would be that, when a fretting seal design is offered, the bidders could be assessed an amount equal to the value of one sleeve every two years. A similar kind of assessment can be conducted for the various types of springs or bellows used as some are more suitable than others for different services.

5.3.3.2 Seal environment

A large bore seal chamber will permit any mechanical seal to last longer. If any bidders do not offer this, they could be assessed the price of an additional spare seal of the type originally quoted.

5.3.3.3 Bearings

In the conventional end suction process pumps, there are two factors in the bearing that can be readily evaluated.

The Size (shaft diameter) and Series (width) of the bearing are directly related to the load carrying capability (LCC) of that bearing. An increase of one Series of the bearing will represent an increase of approximately 50% in the LCC of the bearing. An increase of one Size will allow an increase of approximately 15% in the load carrying capability of the bearing. Specific values can be identified in the bearing manufacturer's engineering data. The larger bearings will last longer.

When one Bidder has one size smaller bearing than the others on both the radial and the thrust side, an assessment of one set of bearings per year will be made at the cost of one set of bearings.

5.3.3.4 Lubricant protection

The L-10 life formula for bearings includes a Life Adjustment Factor which considers the effectiveness of the lubricant. Consequently, whichever lubricant is used, it must be protected against contamination. While many pumps still standardize on the use of lip seals in their bearing housings, most quality pump manufacturers offer (at least as an optional extra) bearing isolators in the housing. These isolators offer protection to the lubricant which is not available from Lip Seals. Any Bidder that does not offer isolators can be assessed the cost of one set of bearings.

5.3.3.5 Rotational speed

From the same formula, it can be seen that the life of the bearing is inversely related to the rotational speed (N) of the pump. Any Bidder operating the pump at the higher speed, can also be assessed the cost of one set of bearings.

5.3.3.6 Shaft slenderness ratio

The Shaft Slenderness Ratio is a simple formula used to identify the strength of the overhung shaft on a centrifugal pump. In this formula, the distance (D) from the centerlines of the impeller and the closest bearing to that impeller is compared to the effective diameter (L) of the shaft at the face of the stuffing box.

A low value of the Shaft Slenderness Ratio will minimize the possibility of deflection and bending, and the amplitude of any vibration that may take place. As a high vibration level and excessive shaft deflection both contribute to premature to premature seal and bearing failure, a more severe evaluation assessment would be appropriate if the pump is in critical continuous duty. For example if any Bidder has a Slenderness Ratio of more than 2.5 times the lowest one, that Bidder would be immediately disqualified from consideration. When a Bidder has a Slenderness ratio of more than twice the lowest value, it could be assessed the cost of a new mechanical seal and shaft.

5.3.3.7 Maintenance hours

It is estimated that the average oil refinery will spend $5,000.00 to replace a mechanical seal. This only covers direct labor and overhead charges. Consequently, consider one seal that costs $750.00 and needs to be changed every year and compare it to another seal which costs $2,000.00 and has to be changed every two years. While the first seal may be lower in purchasing cost, the cost of ownership of the pump over a 10 year period will be ($5,750 × 10) $57,500.00, against ($7,000 × 5) $35,000.00 with the higher priced, but more reliable seal. This displays the importance of including the number of maintenance hours which can be anticipated during the life of the pump.

5.3.3.8 Downtime and lost production

If there is an installed spare ready to run, then that factor can sometimes be ignored. When there is no installed spare, the cost of downtime is frequently so high that many will shortsightedly refuse to recognize and consider it as a real evaluation factor.

5.3.3.9 The results

Some will recognize that we did not consider the existing spares for other pumps in the plant, or the familiarity of operating and maintenance personnel with the new pumps. Such considerations should only be used if there is a very minimal difference in the evaluated value as we are trying to ensure that we will not need a high number of spares and that the amount of work needed on the new pumps will be at an all-time low.

With all these factors taken into consideration, it is not uncommon for the pump which might incur the highest capital cost to be the least expensive to own in only a few years.

When Value Based Purchasing is carried out over an extended period of time, it becomes evident that the initial capital cost of the equipment fades into insignificance when compared to the total cost of ownership.

Stuffing box sealing

6.1 Shaft sealing

For over a hundred years, the leakage of liquids along the shaft from the pump casing was minimized by means of an arrangement of materials, collectively referred to as 'Packing'.

In spite of holding the dubious distinction of being the oldest part of the design of a modern process pump, packed stuffing boxes are still widely used owing to a low initial cost, and because their operation is familiar to plant personnel.

6.2 Packing

Packing is compressed axially into the stuffing box so that it will expand radially and seal against the bore of the stuffing box and onto the pump shaft. As there is no relative motion between the packing and the stuffing box, this function is relatively straight forward. However, with the friction created by the shaft running on the bore of the packing, a certain amount of leakage is essential to lubricate and cool the area.

It must therefore be recognized that the function of packing is not to eliminate leakage from the casing, but rather to restrict the amount of leakage that will occur to about 40 to 60 drops per minute. In spite of this, a certain amount of shaft wear is inevitable and therefore standard practice includes the provision of a protective sleeve over the shaft in this area.

It is interesting to note that the vast majority of wear on the sleeve is a result of the outer one or two rings of packing and these can be relatively easily replaced during a brief shutdown of the pump.

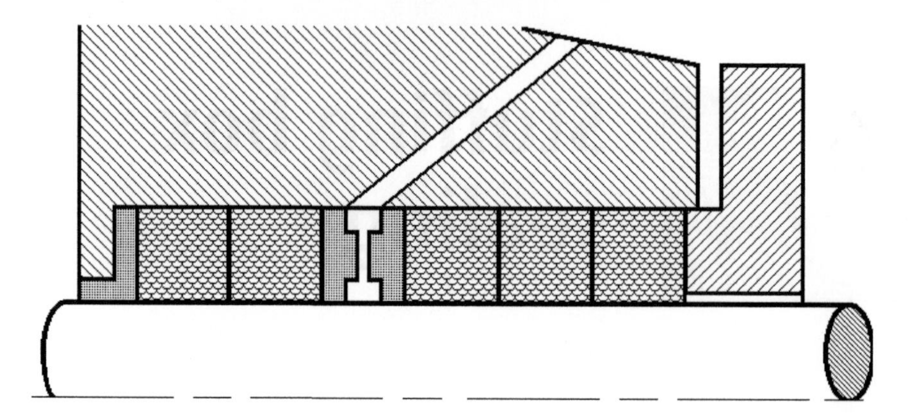

Figure 6.1: Packed stuffing box

When abrasive liquids are being pumped, a lantern ring is installed to accommodate the introduction of a clean sealing liquid that will replace the pumpage in the stuffing box. This sealing liquid is normally introduced at 10 to 15 p.s.i. above the pressure, behind the impeller at the bottom of the stuffing box. This allows the sealing liquid to provide all the cooling and lubrication needed between the shaft and the packing and restricts the entry of the pumpage into the stuffing box area.

The lantern ring is normally inserted about 2 or 3 rings from the bottom of the box where it will connect with the flush port in the stuffing box. However, for extremely abrasive pumpages in a situation where product contamination is acceptable, the lantern ring can be positioned at the bottom of the stuffing box. With appropriate relocation of the flush port, this takes the pressurized sealing water straight into the pumpage and prevents it from damaging the packing.

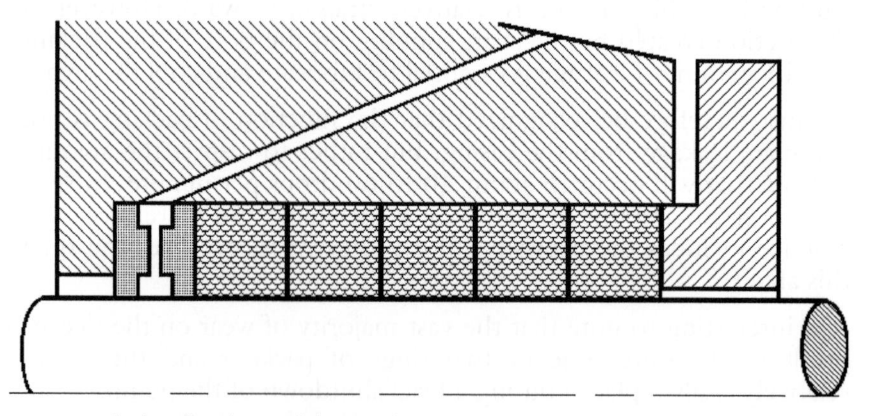

Figure 6.2: Packed stuffing box with bottom lantern ring

It should be noted that this system requires a higher quantity of flush water and most of it will enter the pumpage.

Although the materials from which packing is manufactured have changed considerably since it was first introduced during the nineteenth century, it still provides the same advantages and drawbacks.

Advantages

- It is relatively inexpensive to purchase.
- It is rarely the cause of catastrophic pump failure.
- It can be adjusted or replaced without pump disassembly.
- Most maintenance personnel are accustomed to its use.

Disadvantages

- It lowers pump efficiency.
- Packing and sleeve require regular replacement.
- The packing requires regular adjustment.
- Adjustment requires the touch of an experienced millwright.
- It is required to permit constant leakage.
- It often requires considerable volumes of flush water.

Although constant leakage is required to ensure lubrication between the packing and sleeve, it is now only acceptable if the pumps are handling clean water.

6.3 Mechanical seals

In view of society's increasing awareness of environmental concerns, the leakage required by packing is becoming unacceptable with the more aggressive liquids now common in our industrial processes. Consequently, packing is being replaced by mechanical seals in a growing number of applications.

A mechanical seal operates by having two flat faces running against each other. The rotating face is secured to the pump shaft while the stationary face is held in the gland. As one face is moving while the other is held stationary, his type of seal is referred to as a 'Dynamic' seal.

In a basic seal, there are four possible leak paths which must be secured.

1. Between the two Seal Faces.
2. Between the Rotating Face and the Shaft.

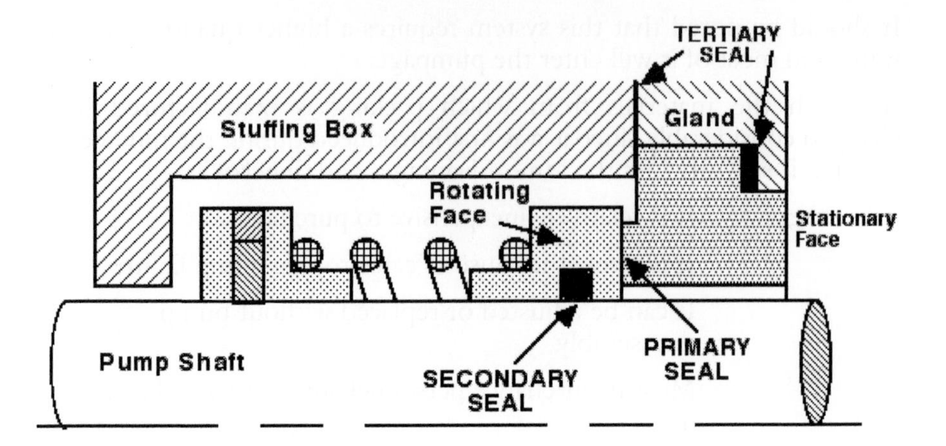

Figure 6.3: Labeled mechanical seal

3. Between the Stationary Face and the Gland.
4. Between the Gland and the Stuffing Box.

The last two seal paths are usually 'Static' seals as there is no relative motion between the two parts. They are frequently referred to jointly as the 'Tertiary Seal', and may consist of a flat gasket or an 'O'-ring in materials compatible with the pumpage.

In the older seal designs, the Secondary Seal under the Rotating Face will move marginally back and forth on the shaft, thus causing fretting corrosion and premature failure. However, in the newer seal designs, the Secondary Seal will be 'Static', thus avoiding fretting corrosion problems on the shaft. This will be discussed in greater detail when we review the 'Fretting Seals' in Chapter 6.3.4.

In normal pump operation, the rotating and stationary faces are held closed by the pressure of the liquid in the stuffing box acting as the closing force. During startup and shutdown, the stuffing box pressure is augmented (and even possibly replaced) by the spring force.

6.3.1 Match the seal to the service

While some liquids are fairly simple to work with, others can be very difficult. It is essential that all the individuals involved (including the seal supplier) are made aware of all the factors that will influence the seal selection. These factors should include the following.

- pressure
- temperature
- corrosiveness
- abrasiveness
- viscosity
- tendency to crystallize

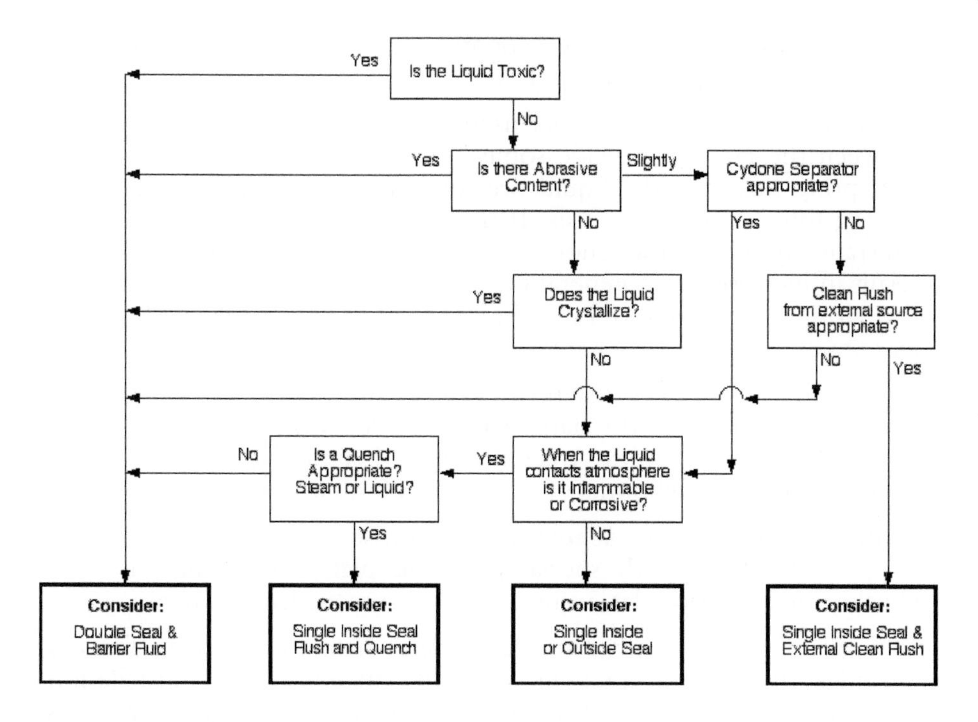

Figure 6.4: Seal selection chart

- toxicity
- operational frequency
- rotational speed
- previous field experience

At this point it is worthwhile to acknowledge the benefit of onsite experience. If there is a history of a particular seal operating well in the same service under similar conditions, then this should take precedence over any other consideration.

It is also in the best interests of the end user to become sufficiently well informed about mechanical seals and their ongoing development, in order to be able to specify the particular seal needed in any application. They can then require all pump suppliers to include that model in their pricing. Alternatively, they can use a method of evaluation such as was suggested in Chapter 5.

6.3.2 The seal faces

Most mechanical seals are designed with a rotating face in a softer material wearing on a harder stationary face. For many years, the most popular combination was the carbon rotating face running on a ceramic stationary. These materials are still in popular use, and have been augmented by adding the option of stainless steel or even harder faces in either tungsten carbide or silicon carbide. Detailed discussions with your local expert will normally identify the best material combination available for your particular application.

Regardless of the face material used, a thin film of liquid must exist between the faces to provide some lubrication. But a combination of the spring load and the liquid pressure in the stuffing box, creates a closing force on the seal faces. If that closing force is too high, it can substantially reduce the amount of liquid between the faces. This will result in increased heat generation and premature wear on the faces. If the closing force is too low, the faces can open easily and permit leakage.

Seal manufacturers are constantly trying to improve the seal design with particular reference to the flatness of the faces. Currently, the industry standard for face flatness tolerance is 2 helium light bands which translates to 23.2 millionths of an inch (0.0000232 inch)

In order to achieve this degree of accuracy, seal faces are machined on Lapping Machines with special polishing plates. The finish is then checked on an optical flat, using a monochromatic light source.

In view of this, it is obvious that careful handling of these faces is essential. It is also apparent that the manufacturers' installation instructions should be carefully followed to ensure that the seal faces are suitably protected and precisely located.

6.3.3 Seal flexibility options

Any axial or radial movement of the shaft will require some flexibility from the spring(s) in order to keep the faces closed.

This flexibility however can only be carried to a certain degree, and the mechanical condition of the pump, including a low value of the slenderness ratio, also plays an important role in the reliability of the seal.

This seal flexibility is usually supplied by

- a single large spring,
- a series of small springs, or
- a bellows arrangement.

The process pumping industry has traditionally used seal designs where the springs were applied to the rotating face. This is known as a 'Rotary Seal' as the springs or bellows rotate with the shaft.

More recent designs apply the springs or bellows to the stationary face of the seal. In fact, it is now quite common to find both stationary and rotating faces of a mechanical seal having some kind of flexible mounting arrangement.

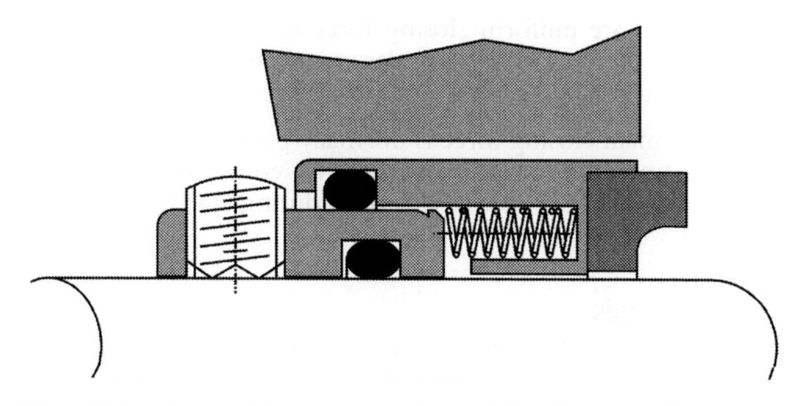

Figure 6.5: Multiple springs seal (Reproduced with permission of AESSEAL plc)

Many seals of an earlier design (as was shown in Figure 6.3) use a large single spring that wraps around the shaft and provides a very strong closing force to the seal faces during the startup of the pump. The installation of the single spring is dependent on the rotation of the shaft to tighten the coil.

The later seal designs opted for a series of smaller springs located around the shaft to provide a more even loading to the seal faces.

As the smaller springs can clog more readily, many seals of this type are designed in such a way that the springs are located out of the pumped fluid entirely.

The most popular design for many aggressive applications is the Metal Bellows Seal which is made from a series of thin metal discs welded together to form a leak tight bellows.

Figure 6.6: Metal bellows seal (Reproduced with permission of A. W. Chesterton Inc.)

This creates a more uniform closing force between the faces, and also eliminates the need for a secondary seal at the seal face, which automatically stops any possible fretting damage.

Although the main closing force is normally provided by the pressure in the stuffing box, the springs and bellows compensate for any shaft movement and keep the seal faces closed during startup and shutdown of the pump.

6.3.4 Fretting seals

A pump shaft will undergo both radial and axial movement for a variety of reasons, including bearing tolerances, end play, vibration and shaft deflection.

In addition, movement within the mechanical seal is also quite normal due to the difficulties in maintaining the two faces absolutely parallel, which is caused by

- Equipment and installation tolerances,
- Thermal growth,
- Pipe strain, or shaft misalignment.

To keep the seal faces together, the springs are constantly adjusting the seal in relation to the moving shaft.

When an elastomer is used between the rotating face and the shaft under these conditions, the elastomer moves back and forth on the shaft. This creates a 'polishing' action which repeatedly removes the protective oxide coating from the corrosion resistant material of the shaft, and eventually creates a groove at that point on the shaft. The groove causes leakage and necessitates repetitive repair or replacement of the shaft.

To combat this problem, the use of the shaft sleeve has been continued as a sacrificial part in the way of the stuffing box.

However, the only lasting solution to the problem of fretting corrosion lies in the elimination of the dynamic seal. Most major seal manufacturers now produce 'non-fretting' seals which protect the pump parts from fretting corrosion. The Bellows Seal shown in Figure 6.6 is a typical example of such a seal.

6.3.5 Balanced or unbalanced seals

The balance of a mechanical seal determines the magnitude of the resultant closing force on the faces. It is created by varying the effective cross-sectional areas of the seal in conjunction with the pressure in the stuffing box.

An Unbalanced Seal exposes the full cross-sectional area of the rotating face to the stuffing box pressure and creates a high closing force between the seal faces.

If a pressure of 50 pounds per square inch is acting on the area of 2 square inches on the reverse side of the rotating face, then the force trying to close the faces is 50 p.s.i., multiplied by 2 square inches, which equals 100 pounds.

Although the nature of the pressure drop across the seal faces is unknown, it is generally considered to be almost linear.

Consequently, the average pressure between the faces will be 25 pounds per square inch, resulting in a force trying to open the seal faces of 25 p.s.i. multiplied by 2 square inches, which equals 50 pounds. Therefore the magnitude of the force trying to close the seal faces is twice that trying to open them. This constitutes an unbalanced seal, the effects of which can include an increased operating temperature and an accelerated wear rate. These conditions would dramatically reduce the seal life in services that were already at a high temperature and/or in an aggressively abrasive liquid.

Balancing a mechanical seal reduces the closing force and extends the life of the seal. This is usually achieved by reducing the effective cross-sectional area of the rotating face by using a stepped shaft or sleeve. However, this is never taken to the point that there is a Net Closing Force of Zero, as it is possible that the condition between the seal faces can become unstable, and they may be blown open by any sudden change.

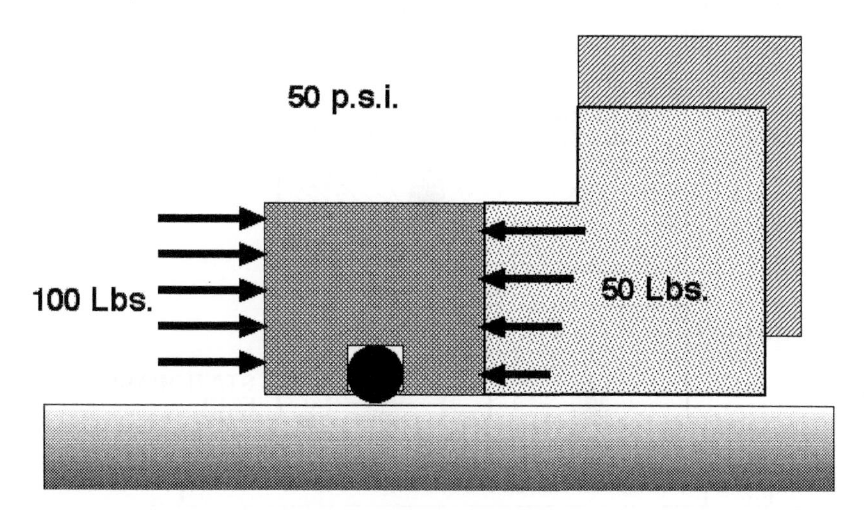

Figure 6.7: Pressures on an unbalanced seal

While the balanced seal may appear to be the answer to all sealing problems, certain services may be better served with the unbalanced seal. For example, some applications may be more interested in security than longevity, and this may translate into a greater desire for a high closing force in the seal selected. Also, when sealing a cold liquid, the increase in operating temperature may be of little concern.

Regardless of any other consideration, a balanced seal is usually recommended when the stuffing box pressure exceeds 50 p.s.i.

6.3.6 Outside seals

The more popular arrangement positions the seal inside the stuffing box. While this requires disassembly of the pump wet end to carry out any maintenance on the seal, there are some significant advantages.

1. Temperature control of the seal area can be achieved by means of a jacketed stuffing box.
2. An aggressive product can be kept out of the stuffing box entirely, by injecting a more acceptable liquid through the flush connections.
3. The seal faces can be kept cleaner by various design elements in the stuffing box that will draw solid particles away from the seal faces.

An Outside Seal reverses the orientation of the stationary face and locates the rotating unit on the shaft outside the stuffing box gland.

The advantages of this arrangement are as follows:

1. Ease of installation.
2. Usually inexpensive.

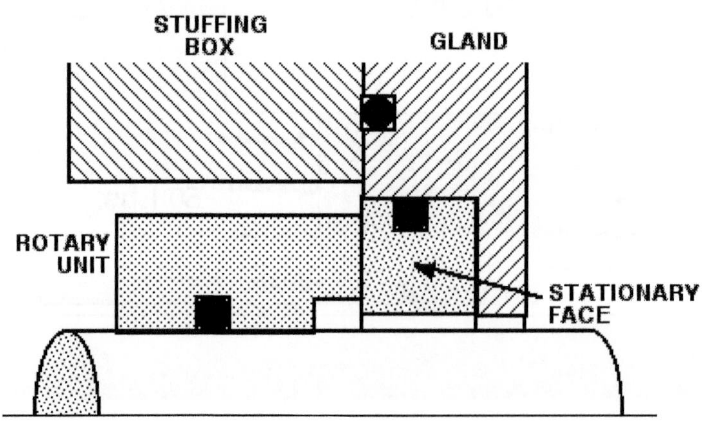

Figure 6.8: Inside seal

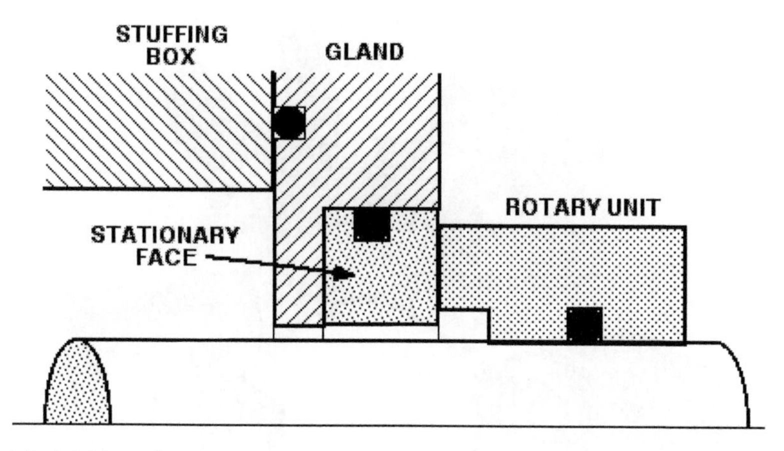

Figure 6.9: Outside seal

3. Permits continual monitoring and cleaning.
4. Can be used when stuffing box is too small.
5. Is less susceptible to shaft deflection difficulties as it is located closer to the bearings.

The major drawback is that centrifugal force will throw any solid particles into the seal faces from the underside of the seal. Consequently, this seal is primarily used with nonabrasive, clean liquids.

6.3.7 Split seals

An important addition to the Outside Seal in recent years is the Split Seal which locates between the stuffing box and the bearing housing. This is a complete assembly designed to eliminate the need to dismantle the pump every time the seal needs to be changed. These seals are gradually being developed to incorporate all the other design criteria discussed (*See* Figure 6.10).

In view of the simplicity of changing the seal with this design, it is important to resist the temptation to merely change the seal when failure occurs, rather than investigating the root cause of such failure.

6.3.8 Cartridge seals

A component seal is one where each part of the seal must be assembled on the pump individually. This requires considerable skill and significant time investment on behalf of the maintenance department.

In addition, there are at least four precautions that must be taken:

1. Establish an accurate operating length.
2. Accurately center the stationary face.

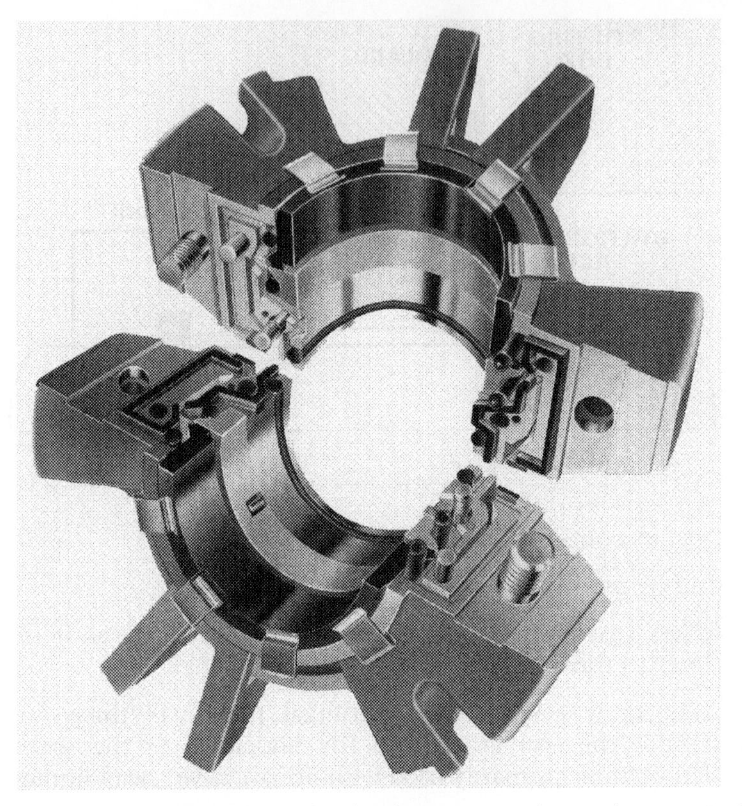

Figure 6.10: Split mechanical seal (Reproduced with permission of A. W. Chesterton Inc.)

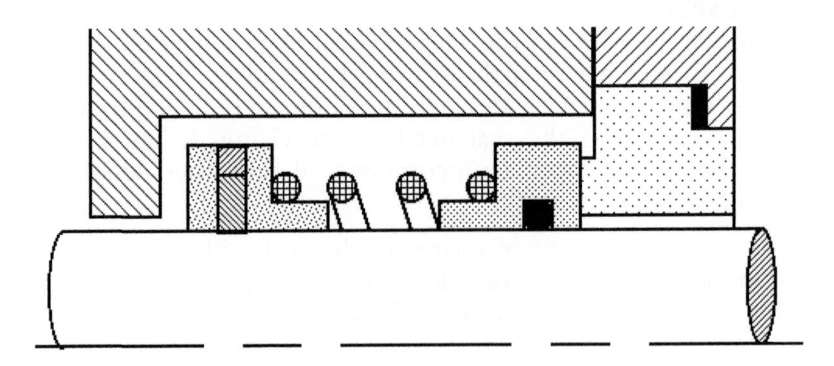

Figure 6.11: Component seal

3. Protect the seal faces.
4. Protect the elastomers.

When using a component type of seal, the manufacturer's critical installation procedures must be followed with the utmost accuracy.

The use of a cartridge seal simplifies these safeguards and eliminates the need to make critical positioning measurements during installation.

The cartridge seal is a completely self-contained assembly which includes all the components of the seal, the gland and the sleeve in one unit.

As it does not require any critical installation measurements, it simplifies the seal installation procedures while simultaneously protecting the faces and elastomers from accidental damage. It also effectively reduces the time spent on maintenance by simplifying seal installation and change-out procedures.

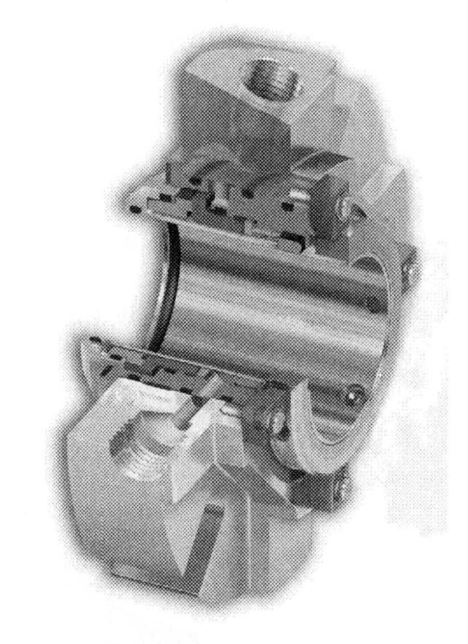

Figure 6.12: Cartridge seal (Reproduced with permission of A. W. Chesterton Inc.)

Cartridge arrangements are available in almost every type of seal on the market, and can therefore eliminate the risk factors and the extra maintenance hours inherent in the use of conventional component seals.

6.3.9 Double seals

A Double Seal is used instead of a Single Seal when a higher degree of leakage protection is desired. It comprises two sets of seal faces combining to increase the security of the environment from the pumpage. They are most frequently used for volatile, toxic, carcinogenic, hazardous and poor lubricating liquids.

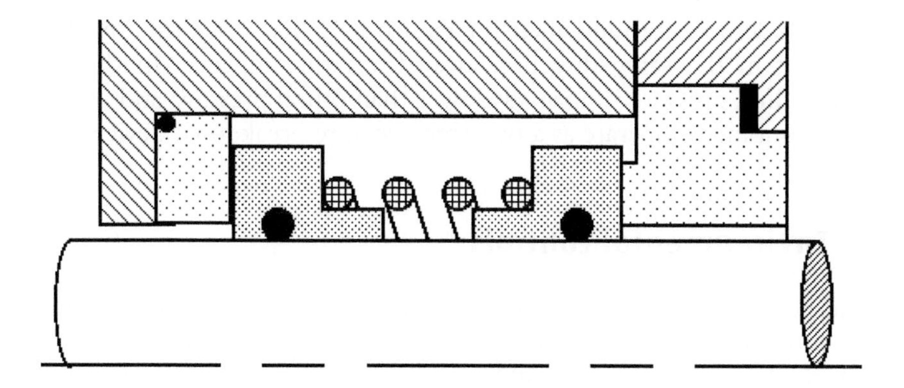

Figure 6.13: Back to back double seal

There are three distinctive arrangements of double seals, all of which require the use of a Barrier Fluid System to maintain a liquid or gas barrier between the two sets of seal faces.

A commonly used low cost double seal arrangement is referred to as the Back to Back Seal. It positions the rotating faces in opposite directions and should always use a Barrier Fluid pressurized to approximately 20 p.s.i. above the stuffing box pressure. This ensures that the inner seal is lubricated at all times by the Barrier Fluid, and also contributes to the closing force on the seal faces.

The more sophisticated Face to Face seal (shown in Figure 6.14), positions the rotating faces pointing towards each other, and they often act on opposite sides of the same stationary face.

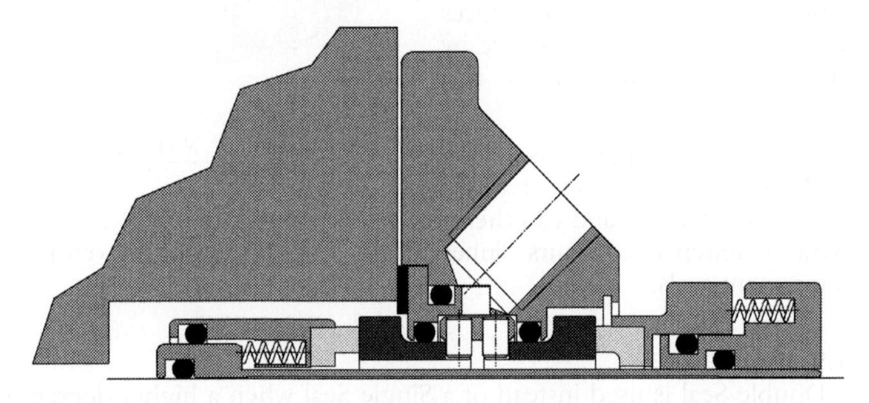

Figure 6.14: Face to face double seal (Reproduced with permission of AESSEAL plc)

This seal can use either a high pressure or a low pressure Barrier Fluid System. It has also been more recently modified as a Gas Seal, using Nitrogen Gas as the inert barrier fluid.

The Tandem seal arrangement in Figure 6.15 has both rotating faces pointing in the same direction away from the impeller. In this, the barrier fluid pressure is normally lower than pump pressure and the two seals combine to operate as a two-step pressure breakdown device.

6.4 Environmental controls

One very important condition necessary for reliable seal operation is the control of the environment in which the seal is located. Even with a low shaft slenderness ratio and a large bore seal chamber, the liquid being pumped may prove difficult to seal without some degree of modification.

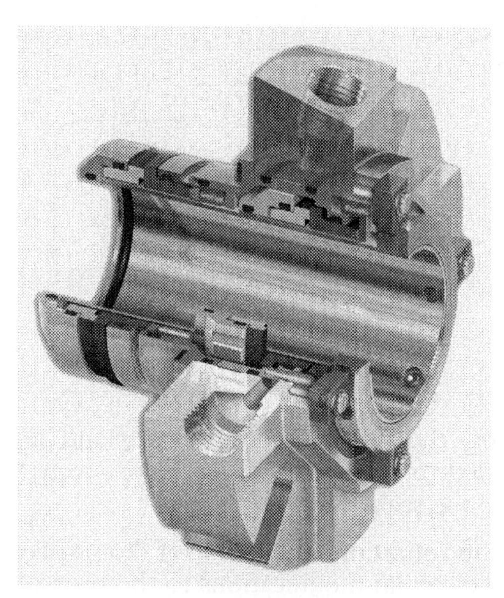

Figure 6.15: Tandem type double seal (Reproduced with permission of A. W. Chesterton Inc.)

Many pumps are equipped with a stuffing box cooling jacket that can be used to cool a high temperature liquid or raise the temperature of an excessively cold product.

6.4.1 Seal flush

Even with a cooling jacket (or when one is not available) an excessively hot liquid may cook the elastomers and distort other parts of the seal, causing premature failure. Similarly, any abrasives in the pumpage can cause rapid wear of the seal faces and also result in premature failure.

In these, and other difficult applications, the product may have to be either diluted or replaced in the stuffing box by an appropriate seal flush, and this can be achieved in a number of ways.

In one method, the flush is brought from a very reliable, external source at a pressure higher than that in the stuffing box. This is frequently used when pumping a dirty product and, in such cases, the flush becomes the main source of lubricant for the seal faces. It will also dilute the liquid and carry any abrasive particles from the seal faces into the pump casing behind the impeller.

In some services the flush would be considered to have contaminated the main process and render this an inappropriate and unacceptable solution.

Other systems use the high pressure process fluid at the discharge nozzle of the pump and recirculate it into the stuffing box.

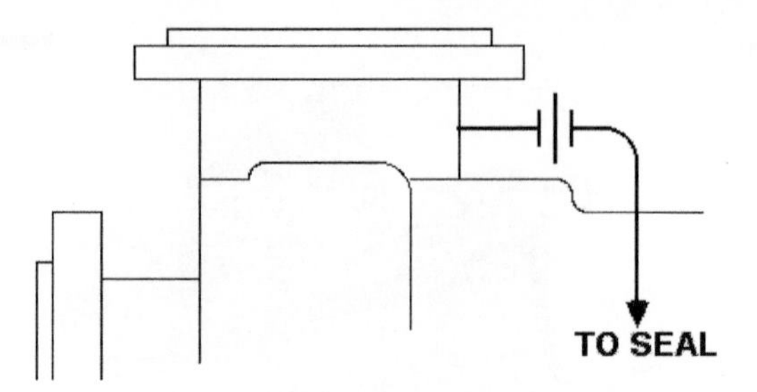

Figure 6.16: Seal recirculation diagram

In these systems, the recirculation pressure and flow rate must be carefully controlled to ensure that it does not create harmful velocities around and onto the seal.

Depending on the condition and nature of the product, it will often be necessary to make certain modifications to the recirculation line, to suit the conditions in the stuffing box. This may involve adding various items such as;

- an orifice,
- a heat exchanger,
- a strainer, or
- a cyclone separator.

These will ensure that the liquid is delivered to the seal, in such a manner as to improve the lubrication, pressure or temperature conditions at the seal faces.

6.4.2 Reverse flush

The reverse flush arrangement is often overlooked. It moves the liquid in the stuffing box to the pump suction. This can be very effective in purging gases from the stuffing box and removing the heat generated by the faces, from the seal area.

In intermittent slurry applications, where the recirculation line is taken from the lowest side of the stuffing box, it is also considered very effective in removing abrasive particles from the seal area during startup (see Figure 6.17).

6.4.3 Seal quench

Unlike the Flush, the Seal Quench never enters the process line. It is designed to remove any leakage from the outside of the seal faces that

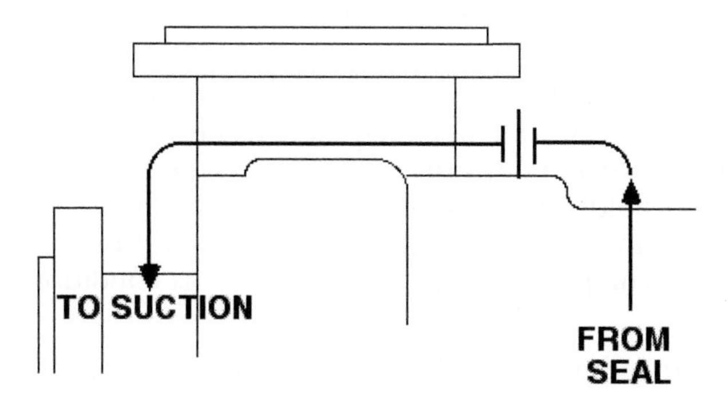

Figure 6.17: Recirculation to suction diagram

would tend to solidify and cause hang-up of the seal faces. It usually incorporates a disaster bushing to keep the quench in the seal area.

Liquid quenches are often used to dissolve and flush out crystalline products, such as from Caustic solutions.

A Steam Quench is frequently used in high temperature applications, to cool the seal faces and prevent coking of leakage residues.

6.4.4 Barrier fluid systems

Barrier fluid systems are required with all types of double seals. They are usually external closed loop systems containing a fluid that is normally different, but compatible, with the process liquid. The system will contain a reservoir which should be as close as possible to the seal.

The design of these systems can vary widely. Some systems will use a pumping ring in the seal while others will use a thermosyphon effect where the difference in fluid temperature between the two legs in the loop, will initiate a continuous flow around the system.

Auxiliary heating or cooling of the fluid is frequently added to the pressure reservoir.

When the barrier fluid system is designed to operate at a pressure lower than the stuffing box pressure, the process liquid will lubricate the inboard set of faces. Any leakage from the inboard seal will flow into the barrier system where it is vented or contained as may be appropriate.

In this arrangement, the outboard seal is cooled and lubricated by the barrier fluid at low pressure until the inboard seal fails. At this point, the barrier fluid system should be isolated to allow process pressure to build and be sealed by the outboard seal which will then carry the full system load until shutdown.

An audible alarm may be designed into the system to alert Operations of the changed condition.

In a high pressure Barrier fluid system, the higher pressure in the Barrier System creates a completely new environment for the inboard seal. Both inboard and outboard seals can now be cooled and lubricated by the barrier fluid. In this condition, a minute amount of barrier fluid will migrate into the process through the inboard seal, so the barrier fluid must be compatible with the process. This condition should also be alarmed.

To further the drive towards zero emissions, the Seal industry has developed Gas Barrier sealing which uses an inert gas, such as nitrogen, to act in place of the Liquid Barrier System.

This requires a special face design on the outer set of faces to ensure its ability to run without liquid lubrication. In addition to ensuring zero product leakage, these Gas Seals also guarantee that the pumped product will not be contaminated by any barrier fluid.

Whether using a liquid or gas barrier system, that system must be dedicated to the specific seal, and alarmed in such a way that any failure of the inner set of faces can be immediately recognized for the appropriate action.

6.5 The seal chamber

Traditionally, the radial clearance between the shaft and stuffing box on the average process pump was sized to accommodate the $^3/_8$ inch square section packing. This resulted in a $1^5/_8$ inch diameter shaft running in a $2^3/_8$ inch bore stuffing box.

When a mechanical seal was introduced into this area, the minimal annular space left available was considered inadequate for reliable operation of the seal and contributed to a high incidence of seal failure. Consequently, the radial clearance between the shaft and the bore of the stuffing box was increased to, at least, $^7/_8$ inch. This has proved to be extremely beneficial in ensuring seal reliability.

By enlarging the bore of the Stuffing Box, the problem identified as seal rub was eliminated. This is a condition where shaft deflection causes the mechanical seal to contact the bore of the stuffing box, resulting in premature seal failure.

Also, the larger volume of the Seal Chamber increased the quantity of liquid around the mechanical seal. This permits greater heat dissipation and allows the seal to operate in a cooler environment.

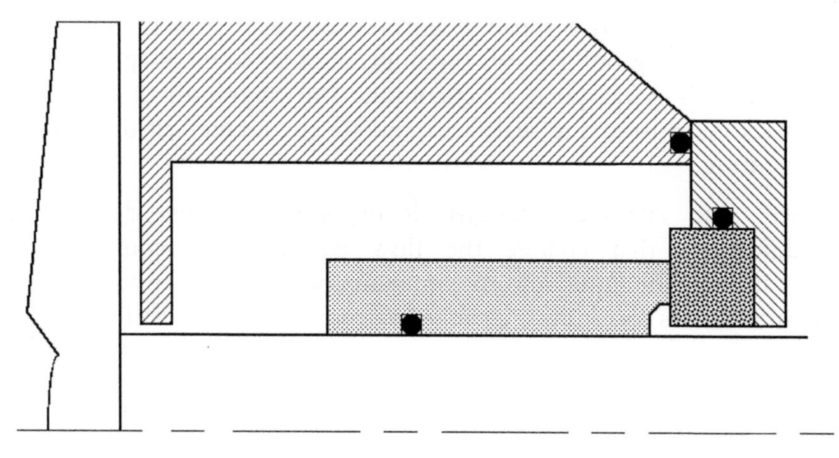

Figure 6.18: Diagram of large bore seal chamber

A number of different seal chamber designs are currently in use. The large cylindrical bore chamber shown in Figure 6.18, is the same design as the stuffing box except that the bore diameter is larger only in the area occupied by the seal, and the traditional close clearance is maintained at the bottom of the chamber.

This design permits temperature or pressure control of the seal chamber. However, it should always be used with a flush arrangement to minimize the possibility of creating air pockets in the chamber.

The through bore chamber eliminates the close clearance at the bottom of the chamber and opens the seal to the full effect of the conditions at the back of the impeller.

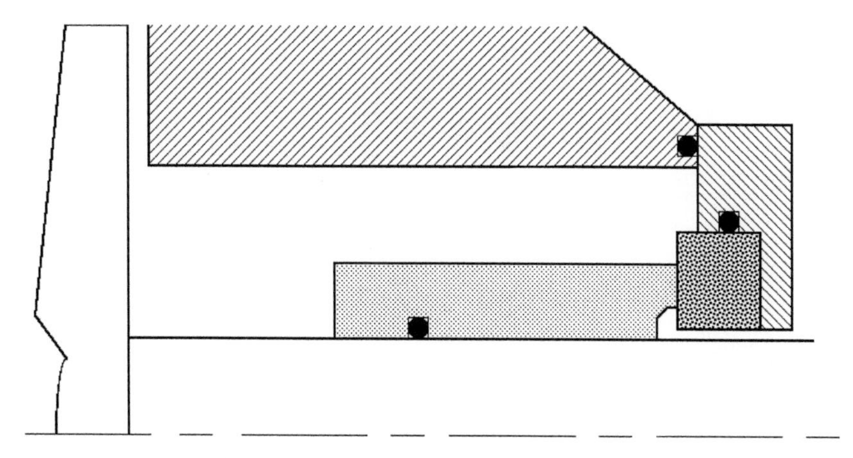

Figure 6.19: Diagram of through bore seal chamber

The tapered bore is a through bore design with a taper bore in the chamber opening out towards the impeller at an angle of approximately 4 degrees. It is designed to encourage the migration of the heavier particles away from the seal faces and to encourage self-draining and self-venting.

Other seal chamber designs incorporate various devices and modifications that change the flow pattern to encourage active circulation of the liquid and to eliminate abrasive particles from the seal chambers, thus providing a cleaner environment for the seal faces.

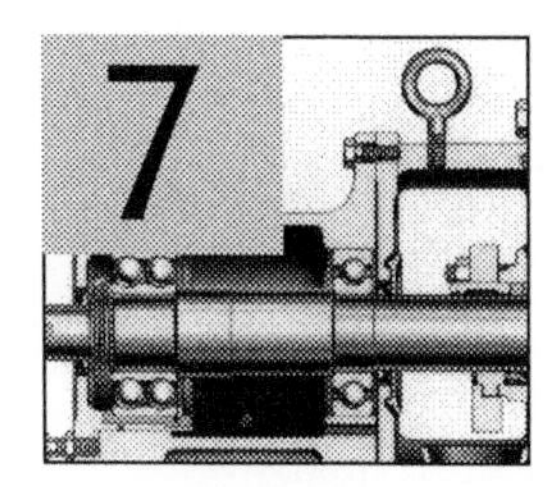

7 Pump bearings

7.1 Pump bearings

The principle function of bearings in centrifugal pumps is to keep the rotating elements in position and in correct alignment with the stationary parts of the pump, while being subjected to radial and thrust loads. The bearings should also permit the shaft to rotate with the least amount of friction to maximize the pump operating efficiency. In addition, they are selected to absorb all the radial and axial loads that are transmitted through the shaft during the different operating modes.

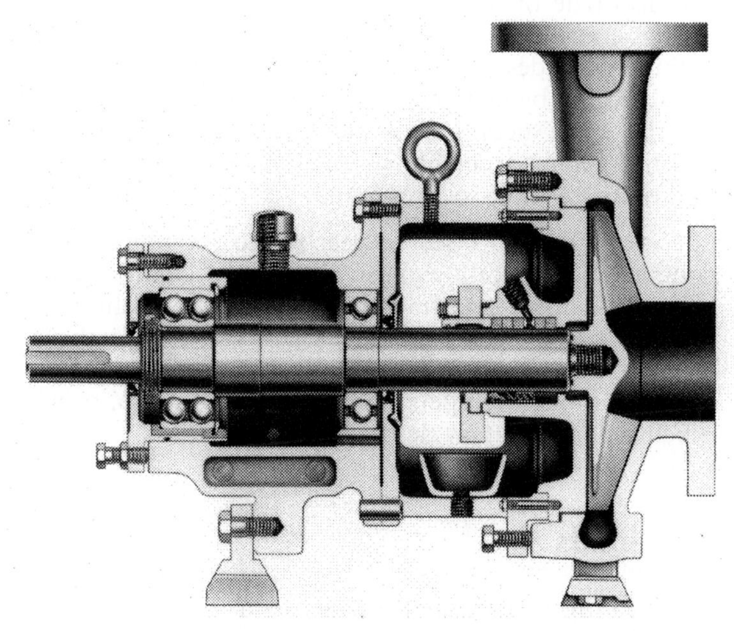

Figure 7.1: End suction pump (Reproduced with permission of Goulds Pumps, ITT Industries)

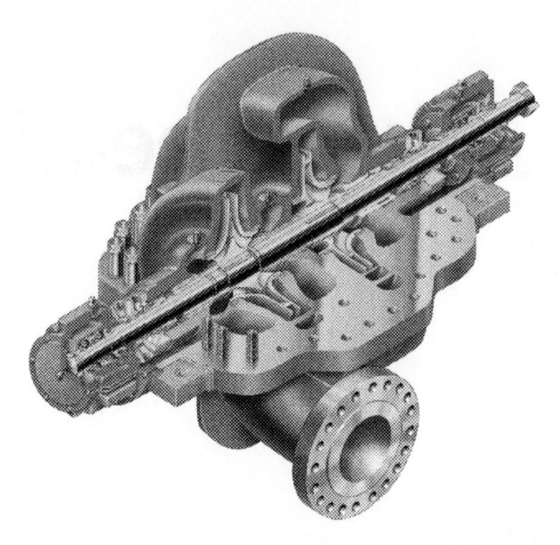

Figure 7.2: Multi-stage pump (Reproduced with permission of Flowserve Corp)

In the horizontal, end suction pump, both bearings are located on one side of the impeller, so that the impeller is mounted on a cantilevered extension of the shaft. Most multistage pumps have one bearing on each end of the shaft with the impellers located between them. In most instances, the radial bearing is situated at the coupling end of the shaft, with the thrust bearing at the outer end.

The same is also true of the double-suction pump in spite of the fact that this style is considered in axial hydraulic balance owing to the inlet flow impacting both sides of the impeller. However, this balance can be detrimentally affected by unequal wear on the wear rings, or the flow of liquid into the two suction eyes may be different because of an improper suction piping arrangement. Consequently, a thrust bearing is still required on the double suction pump.

Bearings are mounted in a housing attached to the pump casing. The housing contains the lubricant and can also provide additional means of controlling the operating temperature of the bearings within acceptable limits. This is frequently achieved with a jacketed area in the housing through which cold water is circulated. It is particularly important in the case of high ambient temperatures around the pump, or when the product is being pumped at an elevated temperature.

7.2 Bearing loads

The life of a bearing is dependent on the loads it must carry and the speed of rotation of the pump. In addition to the physical forces

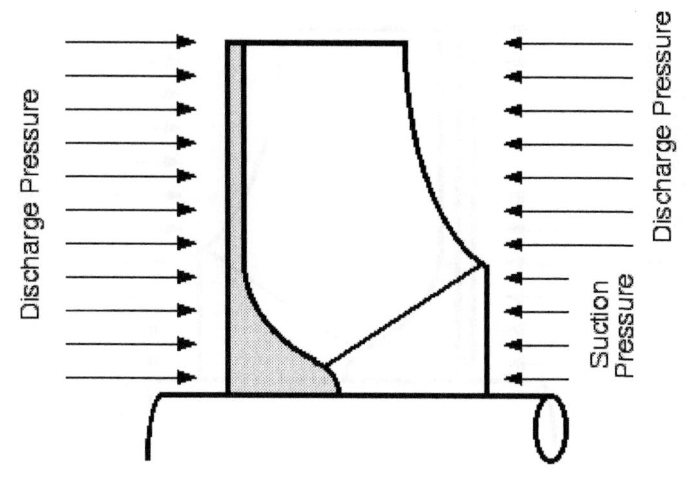

Figure 7.3: Typical impeller thrust loads

involved, the major loads are a result of the hydraulic forces acting in the pump casing at the impeller.

The axial thrust created in a horizontal end suction process pump is predominantly from the hydraulic forces acting on the front and back of the impeller. The magnitude and direction of the resultant axial thrust will depend on the impeller design and flow conditions.

In most instances the resultant axial thrust will be towards the pump suction and is a result of the pressure on the back shroud overcoming

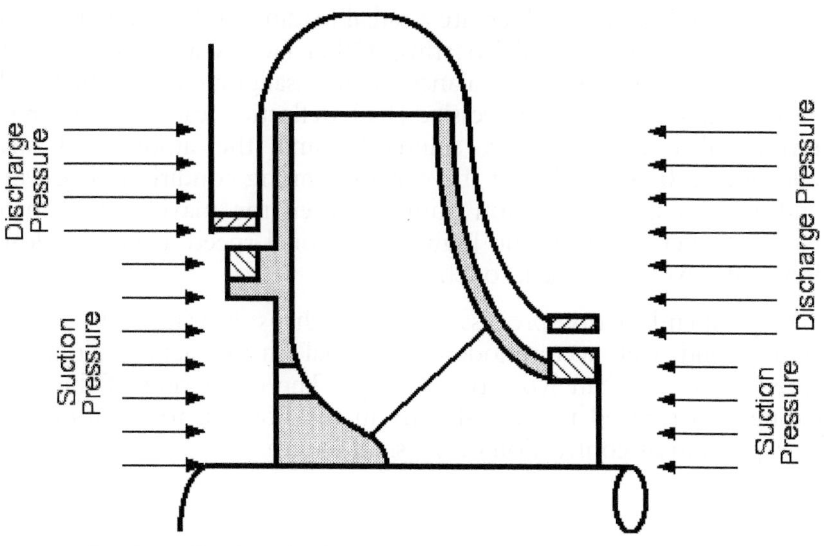

Figure 7.4: Closed impeller thrust loads

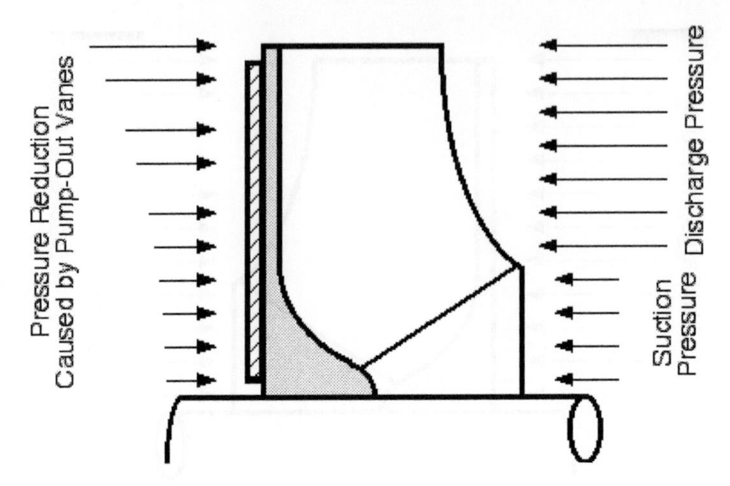

Figure 7.5: Open impeller thrust loads

the pressure on the front side. Closed impeller pump designs often employ back wear rings to reduce the axial thrust. They may also use axial balance holes to connect the high pressure at the rear of the impeller with the suction pressure at the eye of the impeller. This will reduce the axial thrust towards the suction. With high suction pressures, the resultant axial thrust can be reversed.

Open impellers often include pump-out vanes on the reverse side of the impeller to reduce the pressure behind the impeller and thus reduce the resultant axial thrust towards the suction.

The radial force from the impeller can be considered as acting at right angles to the shaft and will create a radial loading on both bearings, but predominantly on the radial bearing. Other factors will also affect the radial loads, such as rotor imbalance, shaft misalignment and the weight of the rotating element. Excessive shaft deflection can have the most detrimental effect. Industry standards limit the amount of shaft deflection to 0.002 inches at the worst operating condition. The worst condition for an end suction pump is when the maximum diameter impeller is being run at the highest rotational speed against a closed discharge valve. Refer back to Chapter 2.5.

In a typical end suction process pump, the thrust bearing is fixed in the housing and will accommodate the axial thrust acting along the centerline of the shaft from the impeller. The outer ring of the radial bearing is permitted to slide slightly in the housing to accommodate any expansion or contraction of the shaft length.

7.3 Ball bearings

While other bearing types are frequently used in special applications, the modern centrifugal pump predominantly depends on the use of the anti-friction ball bearings.

Ball Bearings comprise an inner and outer hardened steel rings separated by a number of steel balls. The inner ring is mounted on the pump shaft while the outer ring is fitted in the housing. The balls are spaced around the bearing by a separator that helps to reduce the amount of friction. These items are manufactured to extremely tight tolerances, and it is essential that the shaft and housing must also be held to the same level of machining tolerances during repair and overhaul. When installing a ball bearing, great care must be taken to ensure accuracy and cleanliness.

Ball Bearings are classified according to the type of loading they are required to accommodate. The size and class of precision are governed by the Anti-Friction Bearing Manufacturing Association (AFBMA) and by the Annular Bearing Engineers Committee (ABEC). Of the five ABEC classifications (1, 3, 5, 7 and 9), Class 1 is standard and Class 9 is high precision. Regardless of the Tolerance Class or the Internal Clearance Class of the bearings used, sound engineering practice requires that the pump original equipment manufacturer's bearing selection be duplicated during repair and overhaul.

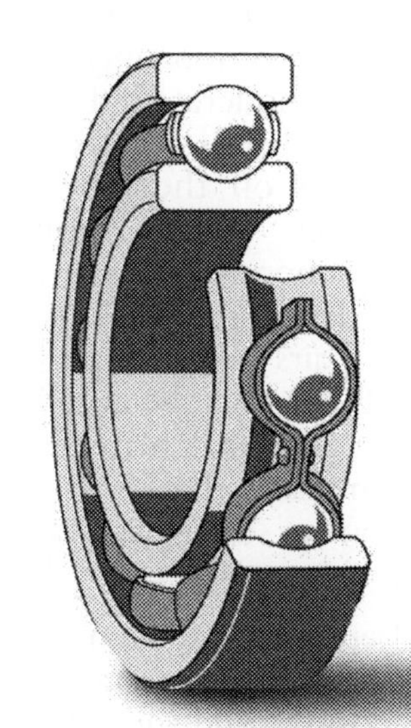

Figure 7.6: Single row deep groove ball bearing (Reproduced with permission of SKF USA Inc.)

Various ball bearings are used on centrifugal pumps, with the most common being the single row, deep groove bearing. In addition to it's ability to handle the radial load, this bearing type is designed with a close contact between the balls and the deep continuous groove in each ring, and is therefore capable of handling an axial thrust load in either direction.

It is also capable of accommodating combined radial and thrust loads. In many small pumps, both the radial bearing and the thrust bearing will be this type of bearing. In the larger sizes, such as a typical ANSI or API pump, it will serve solely as the radial bearing.

The double row, angular contact bearing is a common option as a thrust bearing and is essentially two rows of bearings in a common race. It has a substantial thrust capacity in either direction, and a higher radial capacity due to the two rows of balls. Newer designs and heavier duty pumps tend to use the duplex arrangement of single row angular contact bearings instead of the double row arrangement.

Figure 7.7: Double row bearing (Reproduced with permission of SKF USA Inc.)

A single angular contact bearing is designed to support a heavy thrust load in one direction only. It can also handle a moderate radial load. The contact angle is achieved by a high shoulder on the inner ring and another high shoulder, diametrally opposite, on the outer ring. This design is usually used as matched pairs of single row bearings, but it must be noted that these are usually matched in production so that an even distribution of the load can be achieved without the use of shims.

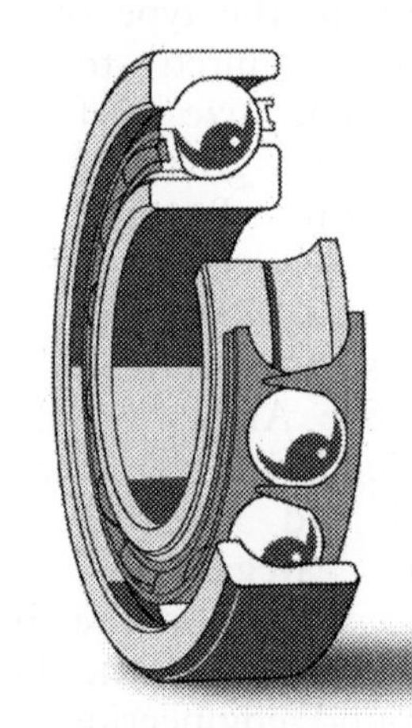

Figure 7.8: Single angular contact bearing (Reproduced with permission of SKF USA Inc.)

Three alternative arrangements of the double angular contact bearings are possible and the load lines must be properly arranged for the anticipated thrust loads.

In the Tandem arrangement, the load lines are parallel and the pair can accommodate axial loads in one direction only with the loading being evenly divided between the two bearings. This arrangement is only used when the design of the pump guarantees the resultant thrust in one direction only.

In the Face to Face arrangement, the load lines converge as they approach the bearing centerline. This arrangement can accommodate axial loads in both directions, but by only one bearing at a time.

It is interesting to note however that, in this arrangement, it is the second bearing that transfer the load from the inner race, through the

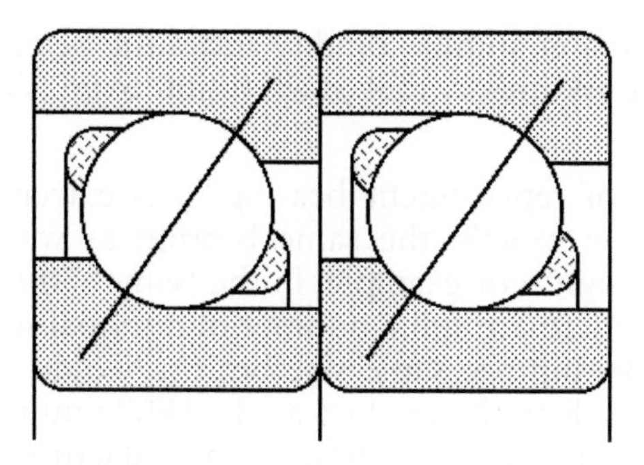

Figure 7.9: Tandem angular contact bearing

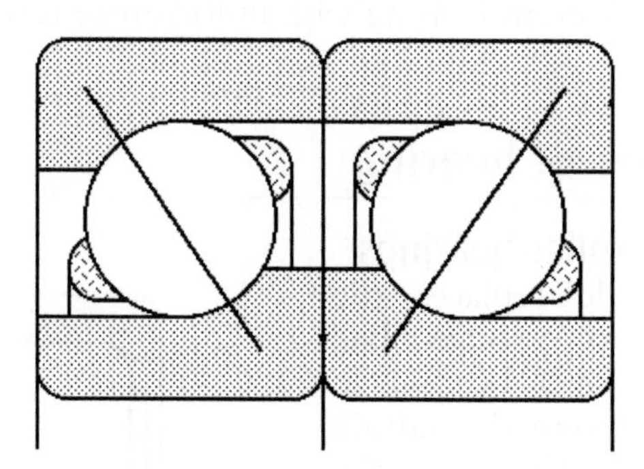

Figure 7.10: Face to face angular contact bearing

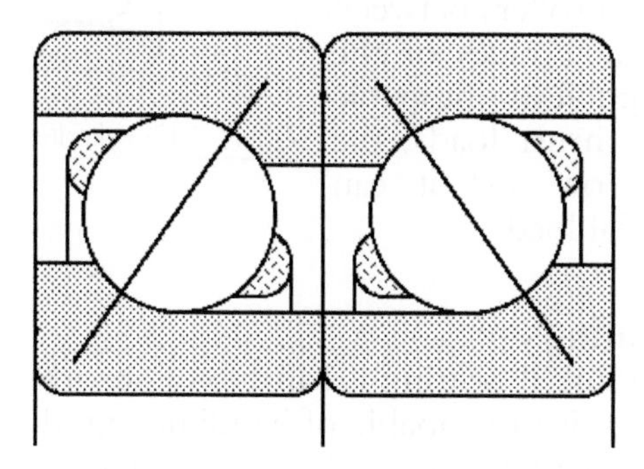

Figure 7.11: Back to back angular contact bearing

ball to the outer race and ultimately onto the bearing housing. This style is not suitable for tilting moments such as may result from shaft misalignment difficulties.

With the Back to Back arrangement, the load lines diverge as they approach the bearing centerline, and this one can also accommodate axial loads in both directions. In this arrangement, it is the first bearing that transfers the load from the inner race, through the ball, to the outer race. It is this arrangement that is normally used in process pumps as it provides a greater angular rigidity and is the better arrangement for accepting tilting moments that may result from shaft misalignment difficulties.

In the purchase of replacement bearing, it is extremely important to ensure you receive exactly the same bearing as was selected for the original pump design. For example, in the 'same' 7000 series of angular contact bearing, there are still a number of optional series and preloads. Other variables within the same size bearing would include cage style and materials as well as the contact angle. While numerical suffixes can be cross-referenced from one bearing manufacturer to another, the alphabetical prefixes and suffixes are frequently different, and you may finish up with a different bearing with unfortunate results.

7.4 Other types of bearing

7.4.1 Cylindrical roller bearings

This bearing style replaces the single row deep groove ball bearing as the radial bearing in some pumps designed outside North America. In this style, the cylindrical roller bearing has a series of cylindrical rollers between the outer and inner races of the bearing. Although it is incapable of handling any thrust loads, the friction is very low and it can support large radial loads.

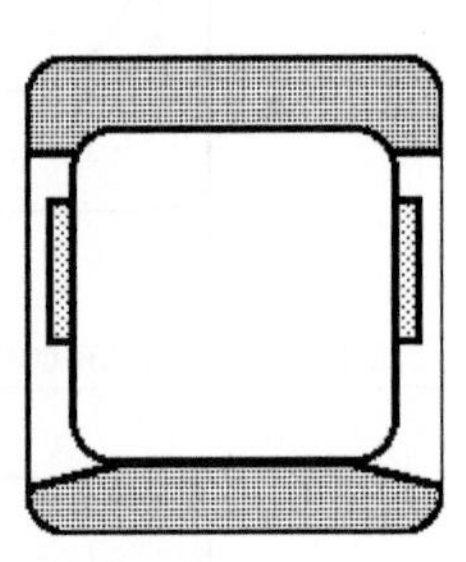

Figure 7.12: Cylindrical roller bearing

7.4.2 Journal bearings

The sleeve type journal bearing is capable of absorbing the radial loads from the shaft, but is not capable of handling any thrust loading. Most frequently used in larger pumps, the sleeve usually contains an internal bearing surface which is most commonly made from a babbitt alloy material. This material minimizes friction while, at the same time, allows any foreign particles to embed themselves in the babbitt and dramatically reduce the potential damage to the shaft.

Good lubrication is critical in obtaining a reasonable life span from journal bearings. To assist in providing a consistent lubricating film between the bearing and the shaft, grooved distribution channels are cut into the bearing surface.

7.5 The total bearing arrangement

To ensure that these bearings provide long-lasting, trouble-free service, it must be recognized that they are only a part of the total bearing arrangement. Other important aspects include the support and the protection of these bearings. Such support includes a strong shaft and housing to minimize the effect of any externally induced stresses or vibration. It also requires accurate machining of the housing and the shaft as well as the correct fits for the bearings.

Protection of the bearings is supplied by the lubricant which is required to separate the rolling elements and the raceway contact surfaces, to minimize the effect of friction, and to prevent corrosion. The selection of the lubricant is a consideration of it's viscosity, and depends on the operating temperature, the bearing size and it's rotational speed. The bearing manufacturer can identify the minimum viscosity required for all these conditions, and the chosen lubricant should provide a higher viscosity than the minimum identified. This reduces the friction losses in the bearing and extends the operating life.

7.6 Oil lubrication

ISO Viscosity Grade denotes the oil Viscosity at 40°C. In other words an ISO VG 100 oil has a viscosity of 100 centistokes at 40°C. As the Viscosity Grade of the oil increases, these oils will provide satisfactory lubrication to the same bearing, but at increasing temperature limitations. Most pump applications will require either an ISO VG 68, or an ISO VG 100 oil to provide a greater safety margin to the bearing lubrication.

Quality mineral oils oxidize at a continuous operating temperature of 100°C, and should be changed every three months or 2000 hours, whichever is the sooner. Synthetic oils however, tend to be more resistant to temperature effects and therefore require less frequent change. As a general guideline, the lubricant should be changed when the maximum contaminant levels of 0.2% of either solids or water are identified.

7.6.1 Static oil lubrication

In a horizontal pump, the oil level should be set at the center of the lowest ball in race. Oil must be able to enter the bearing from both sides, particularly in the case of double row or paired single row bearings where a pumping action from the bearings may repel the oil.

When a sight glass is used in the bearing housing, it tends to provide a more dependable indication of oil level and quality than the constant level oiler. However, it is worthwhile to be aware that the oil level shown in the sight glass may vary slightly when the pump is operating, depending on the direction of shaft rotation and the type of sight glass used.

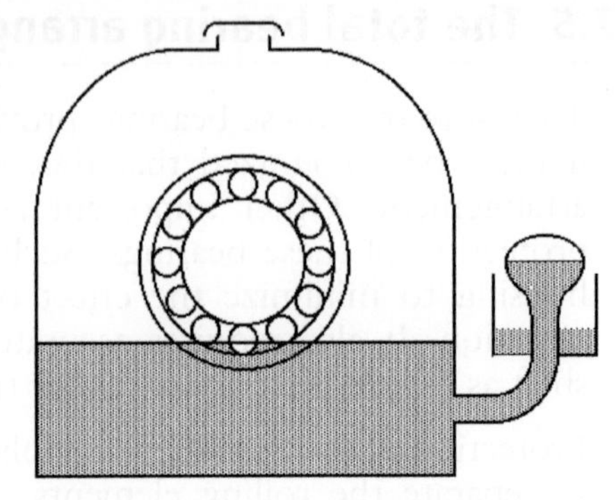

Figure 7.13: Diagram of oiler and level in bearings

7.6.2 Constant level oiler

As it's name implies, the constant level oiler is used to maintain a constant level of the oil in the oil sump or bearing housing. As the oil level is lowered, air enters the oiler and forces the lubricating oil in the oiler to enter the housing. In this way, a constant level of oil is maintained in the bearing housing. It should be noted that, without ongoing maintenance, the constant level oiler has a tendency to clog and thus display a false impression of oil level and quality.

7.6.3 Oil ring lubrication

This arrangement usually comprises a brass ring suspended from the shaft and hanging into the oil sump below the bearing. When rotated with the shaft, the ring draws oil from the sump and throws it to the bearings.

The bore of the ring is generally 1.6 to 2.0 times the diameter of the shaft, and the oil sump should be positioned so that the ring submergence will be approximately 10% to 20% of its diameter. While this system is suitable for high speed operation, it does create wear of the ring components. For full effectiveness of this arrangement, it is also important that the pump centerline be precisely horizontal, as any tipping will cause the ring to run

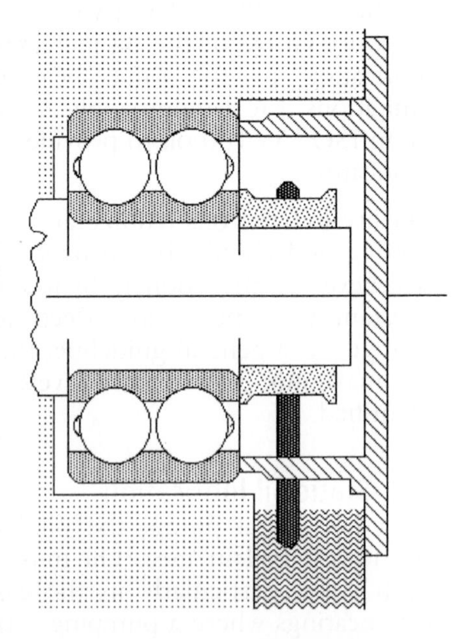

Figure 7.14: Diagram of oil ring lubrication

on its edge, thereby reducing its ability to bring the oil up from the sump.

7.6.4 Oil mist lubrication

An oil mist is a collection of atomized oil droplets which are sprayed into the bearings by compressed air at a pressure just above atmospheric. It is used to fill the bearing housing as much as possible with oil in order to minimize the entry of contaminants. In a Purge Oil Mist system a static oil bath is contained in the housing as well as the oil mist spray.

In a Pure Oil Mist system, the oil bath is eliminated and the spray of oil mist is the only form of lubrication used. This results in a minimum

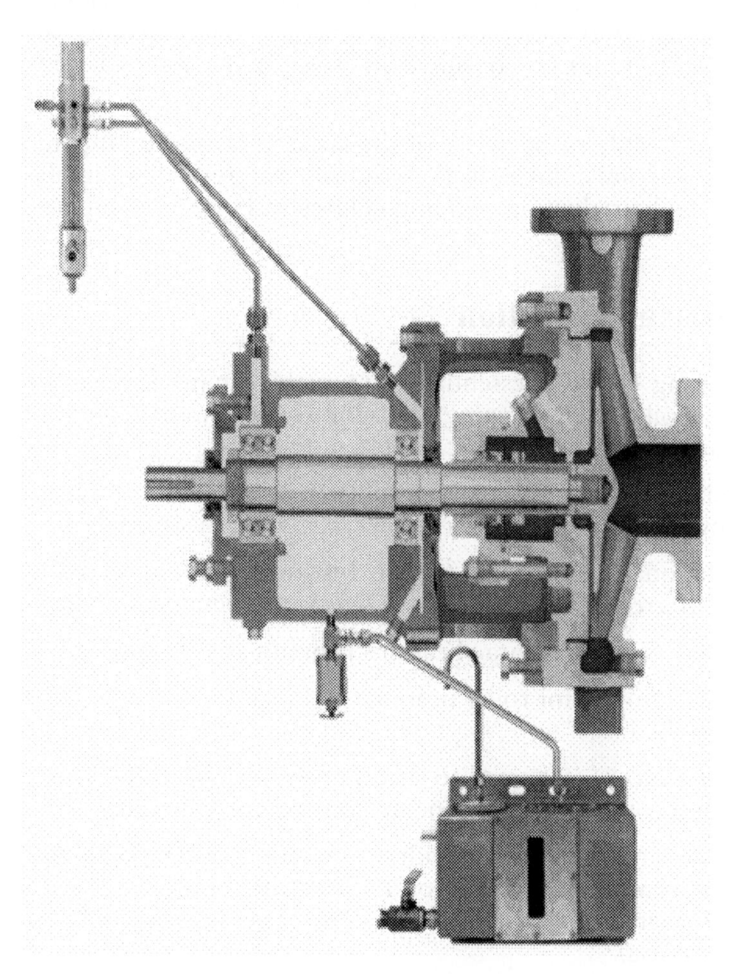

Figure 7.15: Diagram of oil mist system (Reproduced with permission of Lubrication Systems Company)

amount of bearing friction and a longer bearing life. It is worthwhile noting however that, in the case of an extended interruption in the flow of the oil mist, the bearings may be susceptible to premature failure.

The mist can be applied into the housing or directly at the bearings themselves. The latter is generally used on bearings with a high thrust load, or when the rotational speed of the shaft (in rpm), multiplied by the bearing mean diameter (in mm.) is greater than 300,000. The oil mist must be vented to the ambient atmosphere on the bearing side opposite the oil mist application to allow an effective oil mist flow across the bearing.

Reliable operation of an oil mist system is dependent on the correct physical layout and installation of the system. Also, special dewaxed oils are recommended for use in oil mist systems to prevent clogging of the mist fittings and other related problems. Manufacturers of oil mist equipment should be consulted for specific recommendations in both these areas.

It is worthwhile to be aware of the fact that a very high percentage of users of oil mist lubrication systems have recorded significant cost savings in the reduction of pump failures and direct maintenance costs.

7.7 Grease lubrication

Lubricating grease is essentially a soap thickening agent in mineral or synthetic oil. When selecting the right grease, the base oil must be able to satisfy the bearing's lubrication requirements.

The mixing of different types of lubricating grease is not recommended as many of their contents and preservatives are incompatible.

The National Lubricating Grease Institute suggests different consistencies of grease for various pump sizes and operating conditions.

- NLGI No 1 – for large bearings running at slow speeds
- NLGI No 2 – for roller bearings and medium-to-large size ball bearings
- NLGI No 3 – for small-to-medium size ball bearings; also in vertical pumps, or pumps experiencing considerable vibration.

The following equation provides a general calculation of the interval between bearing regreasings on a horizontal pump, and is based on a bearing operating temperature of 70°C (158°F), and using an average quality grease.

$$T = K \left[\frac{14.0 \times 106}{n \times \sqrt{d}}\right] - 4d$$

T = Grease service life (hours)
K = Factor for bearing type (10 for ball brgs.)
n = Pump rotational speed
d = Bearing bore (mm)

In spite of this calculation, regreasing intervals will have to be more frequent than shown above when the bearings are operating at higher temperatures, the bearing axis is vertical, or if (as shown later in this Chapter) contamination is present. Conversely, the calculated interval can be increased when the bearings are operating at lower temperatures or if a higher quality and/or temperature greases are used.

A good rule of thumb is to schedule regreasing every 2,000 operating hours or every 3 months, whichever comes first. It should be noted that regreasing should take place only when the pump is not running.

7.7.1 Shielded bearings

A single shield on a bearing can control the amount of grease entering that bearing. The bearing should be positioned in the pump with the shield located on the side towards the grease supply. This allows sufficient grease to pass through the shield clearance to lubricate the bearing and any excess will pass through the bearing.

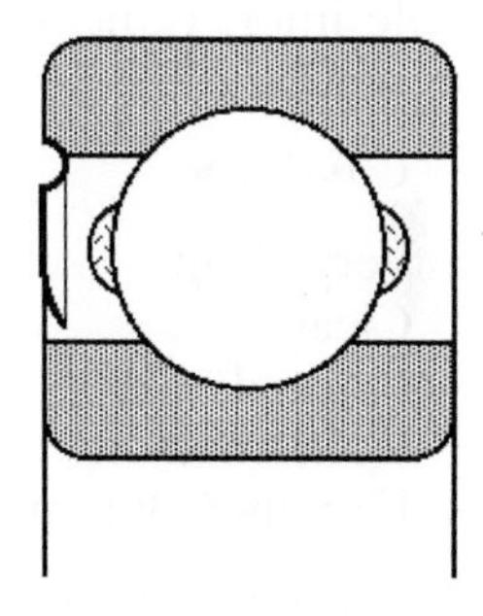

Figure 7.16: Shielded bearing

This arrangement must be used with some caution as some of the newer shield designs require higher pressures to fully inject the grease, yet too much pressure can collapse the shield into the cage.

7.8 Bearing life

Many pump specifications require a bearing life stated in terms of an 'L-10' fatigue life. The L-10 life is that which 90% of a sufficiently large group of apparently identical bearings can be expected to attain, and can be calculated as shown. In other words, only 10% of these bearings will fail within that number of operating hours.

$$L\text{-}10 = A_{23} \left[\frac{C}{P}\right]^{p} \times \frac{1{,}000{,}000}{60 \times n}$$

a_{23} = Lubrication effectiveness factor
C = Basic dynamic Load rating
P = Equivalent dynamic bearing load
p = Exponent of the lofe equation (3 for ball brgs.)
n = Pump rotational speed

A number of adjustment factors in the calculation of operating hours of bearings have been introduced over the years. One of which is based on the relationship between the viscosity required for adequate lubrication and the actual viscosity used. By selecting a lubricant with a sufficiently high viscosity, the bearing life can be increased by a factor as high as 2.5. However, it can also reduce dramatically.

Another adjustment factor that has been suggested involves the effect of varying degrees of contamination of the lubricant. This implies that we will lose 40% of our bearing life under what is referred to as 'Normal' operating conditions, which seems to indicate that considerable improvement is possible on the cleanliness of our bearing environments.

Condition		*Adjustment Factor*
Very clean	=	1.0
Clean	=	0.8
Normal	=	0.6
Contaminated	=	0.5 – 0.1
Heavily Contaminated	=	0

It must be noted that these factors are intended to be combined with the other relevant data on lubricant viscosity before being applied directly to the bearing life formula.

Where water contamination is a problem, it is generally accepted that a mere 0.002% water content in wet oil results in a loss of 48% of bearing life, and 3% of water content results in a loss of 78% of bearing life.

All of this underscores the necessity for protecting the quality of the lubricant used to ensure bearing reliability in centrifugal pumps.

7.9 Lubricant protection

A variety of methods are used to keep the contaminants out of the lubricant in pumping equipment.

7.9.1 Sealed bearings

When available in deep groove and double row styles, sealed bearings may be appropriate. These bearings are fitted with two seals and are not regreasable. They are considered 'sealed for life' (the life of the bearing). In these conditions the bearing cavity is filled to approximately 25% to 35% with grease. This offers a good guide to those who have a tendency to over-grease bearings. However, the life expectancy of the bearings still depends on the L-10 fatigue life of the bearing and the quality of the grease.

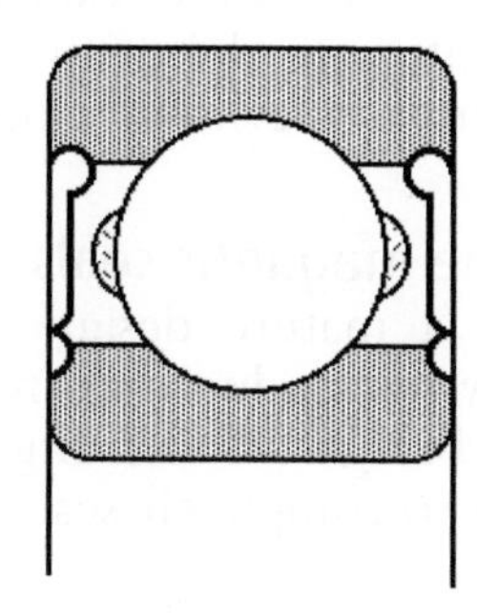

Figure 7.17: Sealed bearing

7.9.2 The lip seal

The most commonly used device to keep contaminants out of the bearing housing is the lip seal, in spite of the fact that it has long been proved to be inappropriate for it's intended function in a centrifugal process pump. It should be noted for the purposes of clarification that, the function of the seal in the bearing housing is to keep contaminants out of the housing, and not to keep the lubricant in the housing.

The lip seal is designed for the equivalent of an L-10 life of 1,000 operating hours. In a pump operating on continuous duty, that translates into a period of less than 6 weeks. This life span is also dependent on the lip seal being well lubricated, yet most process pumps

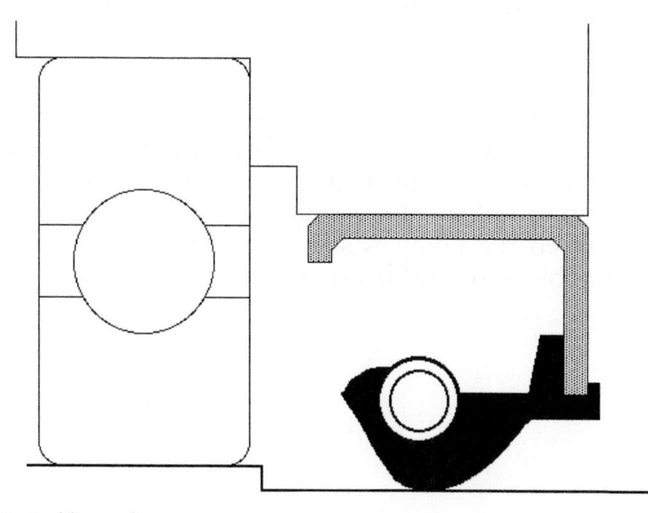

Figure 7.18: Typical lip seal

have little or no lubricant in the immediate area of the lip seals, thus contributing to their early failure. In addition, the lip seal is a contact seal that will inevitably groove the shaft.

7.9.3 The magnetic seals

This is a mature design whose use in centrifugal pumps has been recently reintroduced as an option to the lip seal. A current double seal design brings the rotating faces into full contact with the stationary magnet to completely seal off the housing.

7.9.4 Bearing isolator

A less damaging option is a non-contacting labyrinth seal or bearing isolator, and these are available in a variety of configurations. All of them have a rotor secured to the shaft by 'O'-rings that drive the rotor without inflicting any fretting damage. The stator collects the liquid centrifuged away from the rotor, and drains it away from the lubricant to the outside of the housing.

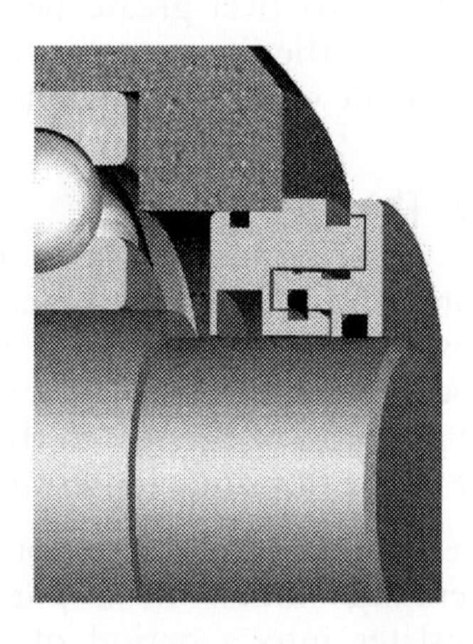

Figure 7.19: Bearing isolator (Reproduced with permission of Inpro/Seal Company)

Most of these seals are designed to operate on horizontal shafts with a drain hole that is required to be installed at the lowest (or six o'clock) position.

Earlier designs exposed the inside of the bearing housing to external ambient conditions when the pump was not running. While this was acceptable in many instances, in areas of high humidity it permitted the ingress of sufficient moisture to perpetuate the problem. To combat this, some bearing isolator designs are now available that will seal off the housing when the pump is stationary. Under these conditions it would be advisable to seal off the traditional breather cap.

8 Special applications

8.1 Slurry pumping

Slurry is a mixture of solid particles suspended in a liquid that is usually water. The variety of solids that are handled in slurry form covers an extraordinary wide range of products and waste material.

One of the key elements in slurry pumping is the size and nature of the solids being transported by the water and the nature of the abrasive wear it causes. As wear is a function of velocity, the pumps usually operate at 1200 rpm or slower.

The centrifugal pumps used in slurry handling are basically conventional water pumps modified in a variety of ways to handle the particular solids. While some of these modifications are minimal, some of them are quite extensive. The difference depends on the size and nature of the solids being handled.

Although the emphasis on a slurry pump tends to be on the size and percentage of solids to be moved, it should be remembered that many of these slurry applications are in services where corrosion resistance is also a factor. In such cases, material selection for corrosion resistance needs to be combined with the pump style selection.

8.1.1 Industrial slurries

In general industry where water run-off in the plant may drain to a central sump and is then pumped out to a collection tank, the solids in the water usually represent plant debris and tend to be quite small. As they also represent a low percentage of the total volume, the slurry can usually be handled by a conventional centrifugal sump pump. The specific pump design in these services can be quite varied, but the traditional vertical submerged suction sump pump and the submersible pump are both widely used.

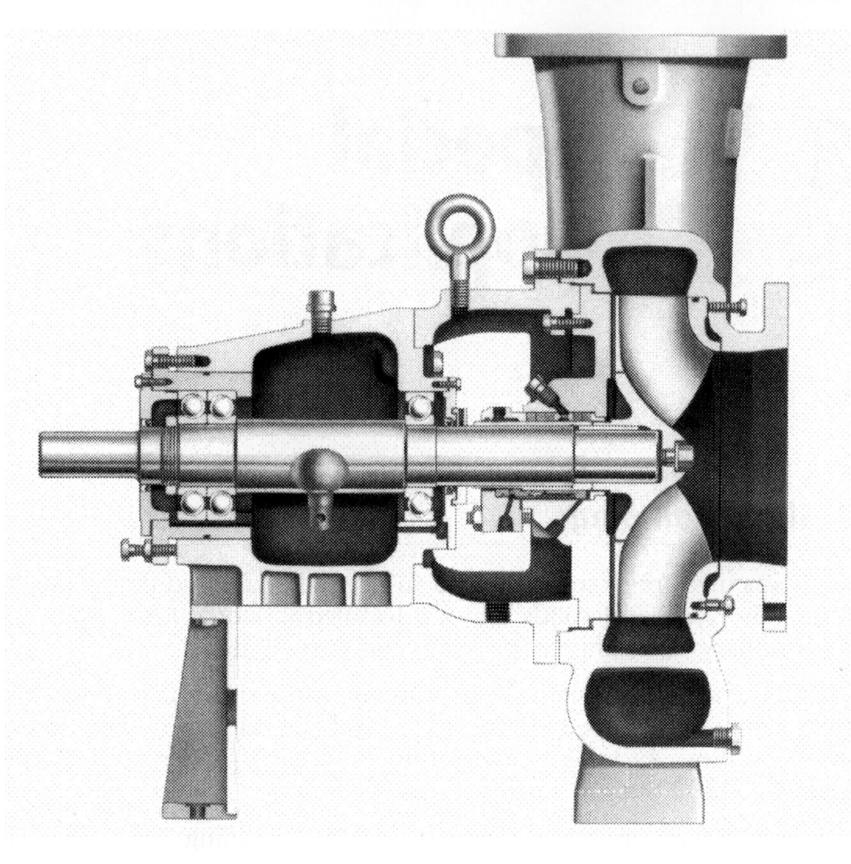

Figure 8.1: Solids handling pump with wear plate (Reproduced with permission of Goulds Pumps, ITT Industries)

For special industrial applications a variety of pump designs have been developed. A popular option to the large closed impeller in some industries is the open impeller that operates with a tight clearance against a casing fitted with sacrificial wear plate as shown in Figure 8.1.

8.1.2 Municipal waste

In municipal waste management applications, solids handling pumps as shown in Figure 8.2, are designed with the capability of handling specified spherical diameters. A few models of such pumps will have an open impeller while many will be designed with closed impellers having the necessary clearance between the vanes of the impeller and also between the shrouds.

For example, a 4 inch pump would have the ability to pass a 4 inch sphere through the impeller. The same pump would have at least a 4 inch diameter suction nozzle and a 4 inch diameter discharge outlet.

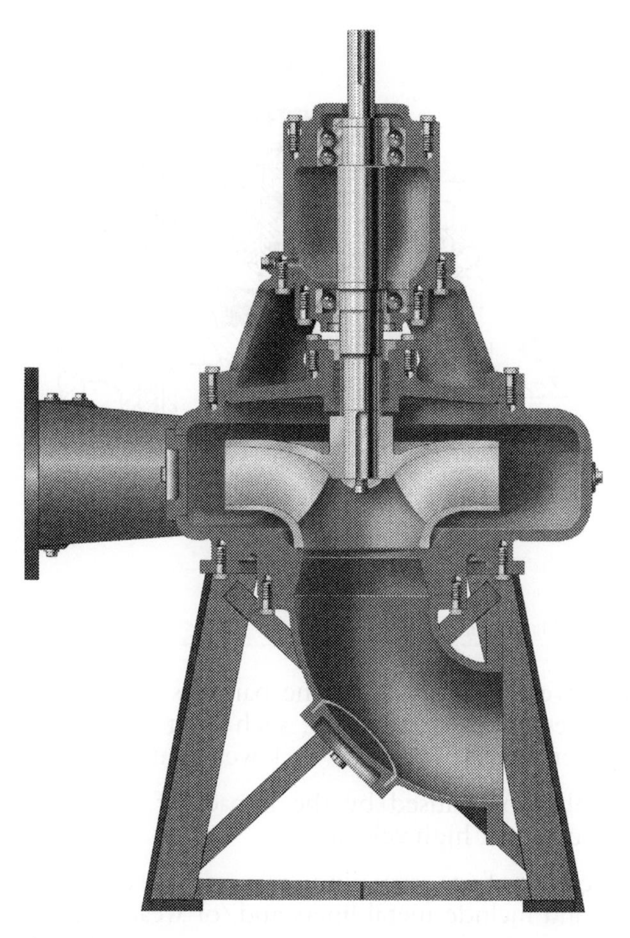

Figure 8.2: Typical non-clog pump (Reproduced with permission of Flowserve Corporation)

This type of pump defies the normal design custom on a centrifugal pump of having the discharge being one size smaller than the suction.

In spite of the focus on the solids size in the selection and purchase of pumps for municipal waste, the major problem in these pumping applications tends to be the stringy material. This material can invade the eye of the impeller, wrap itself around the shaft nut, and eventually clog the pump.

8.1.3 Pipelines and mines

These applications usually require very large pumps that are subjected to high levels of abrasive wear of different types.

- Gouging abrasion occurs when coarse, angular particles tear fragments of the wearing surface.

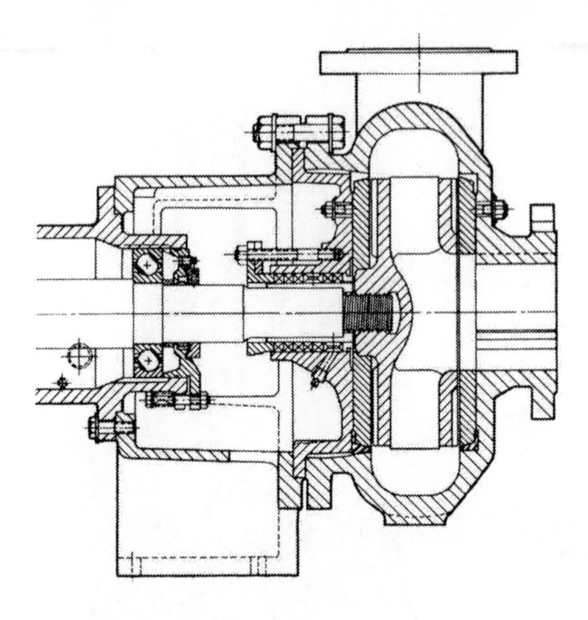

Figure 8.3: Hard metal lined pump

- Grinding wear as a result of fine particles being crushed between two surfaces in close proximity, such as at the clearances between the impeller and the front and back wear plates, or the casing itself.
- Erosion abrasion caused by the impact of solid particles on the wearing surface at high velocity.

Tough materials of construction are necessary in most of these applications and include metal liners and/or wear plates with Ni-Hard

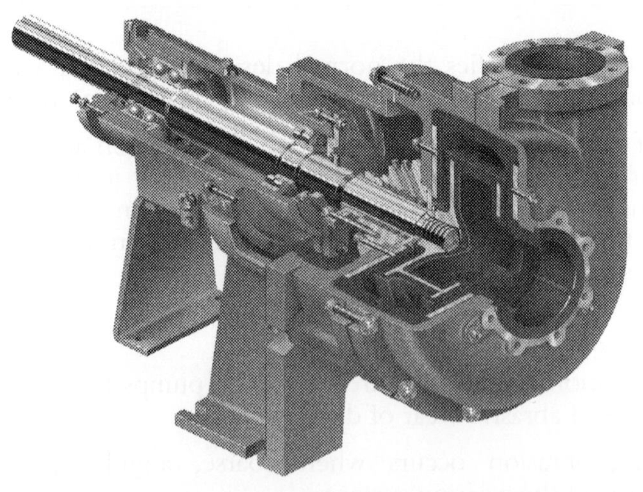

Figure 8.4: Rubber lined pump (Reproduced with permission of Flowserve Corp.)

and heat-treated high chrome iron. Natural rubber and other elastomers are also frequently used as a liner for abrasive services as long as they are chemically compatible with the slurry.

It is worth noting that hard metal and rubber impellers cannot be machined to the necessary diameter in order to meet the required operating conditions. Consequently, they have traditionally been belt-driven and a change in operation is achieved by a change in sheave ratio to give the rotational speed required.

8.1.4 Recessed impeller vortex pumps

In most slurry, the water tends to be used simply as a mode of transport for the solid particles which usually impart the damage to the pump. This is when the special materials are required for the casing and impeller to resist such attack as discussed above.

However, there are also those conditions where the solids part of the slurry is the important part and needs to be protected from the damaging impact within the pump. For such applications, a popular option is the recessed impeller (or vortex) design shown in Figure 8.5.

This style locates the impeller in a recessed position in the casing out of the normal flow pattern in the pump. The impeller develops a vortex in the fluid inside the pump casing so that most of the solids never touch the impeller. While this tends to reduce the wear on the impeller, it also minimizes any damage to the solids.

Figure 8.5: Vortex pump (Reproduced with permission of Fairbanks Morse Pumps, member of the Pentair Pump Group)

8.1.5 Diaphragm pump

One of the original diaphragm pump designs actuates a single large diaphragm in a horizontal casing by means of a spring or a linkage mechanism. Many of these are still in service in the Municipal markets and are used to pump heavy sludges and debris laden waste from manholes and catch basins.

The newer designs include the Air Operated Double Diaphragm (AODD) pump that has become an industrial standby in spite of the fact that is limited to low pressure applications and provides a pulsating flow.

8.1.6 Progressive cavity pump

The Progressive Cavity pump has recently developed a place in the slurry market for delivering smooth, non-pulsating flows. These pumps are particularly effective when pumping thicker sludges with limited amounts of small solid particles. They usually operate at less than 300 rpm to minimize wear. See Chapter 9.4.5 for further details on this pump style.

8.1.7 Solids and slurries – useful formulae

The formula for specific gravity of a solids-liquids mixture or slurry is as follows:

$$S_m = \frac{S_s \times S_l}{S_s = C_w (S_l - S_m)}$$

where
S_m = specific gravity of mixture or slurry
S_l = specific gravity of liquid phase
S_s = specific gravity of solids phase
C_w = concentration of solids by weight
C_v = concentration of solids by volume

Example: If the liquid has a specific gravity of 1.2, and the concentration of solids by weight is 35%, with the solids having a specific gravity of 2.2, then the following formula will apply.

$$S_m = \frac{2.2 \times 1.2}{2.2 + 0.35\,(1.2 - 2.2)} = 1.427$$

Where pumps are to be applied to mixtures which are both corrosive and abrasive, the predominant factor causing wear should be identified and the materials of construction selected accordingly. This often results in a compromise and in many cases can only be decided as a result of test or operational experience.

For any slurry pump application, a complete description of the mixture components is required in order to select the correct type of pump and materials of construction.

$$C_w = \frac{\text{Weight of dry solids}}{\text{Weight of dry solids} + \text{Weight of liquid phase}}$$

$$C_v = \frac{\text{Volume of dry solids}}{\text{Volume of dry solids} + \text{Volume of liquid phase}}$$

A nomograph for the relationship of concentration to specific gravity of dry solids in water is shown in Figure 8.6.

Slurry flow requirements can be determined from the following expression.

$$Q_m = \frac{4 \times \text{dry solids (in tons per hour)}}{C_w \times S_m}$$

where Q_m = slurry flow in USGPM

1 ton = 2,0001lbs

Example: 2,400 tons of dry solids are processed in 24 hours in water with a specific gravity of 1.0, and the concentration of solids by weight is 30% with the solids having a specific gravity of 2.7 then the following will apply.

$$S_m = \frac{2.7 \times 1.0}{2.7 + 0.3\,(1.0 - 2.7)} = 1.23$$

$$Q_m = \frac{4 \times 100}{0.3 \times 1.23} = 1{,}084 \text{ USGPM}$$

8.1.8 Abrasive wear

Wear increases rapidly when the particle hardness exceeds that of the metal surfaces being abraded. Always select metals with a higher relative hardness to that of the particle hardness. There is little to be gained by increasing the hardness of the metal unless it can be made to exceed that of the particles. The effective abrasion resistance of any metal will depend on the position on the Mohs or Knoop hardness scale.

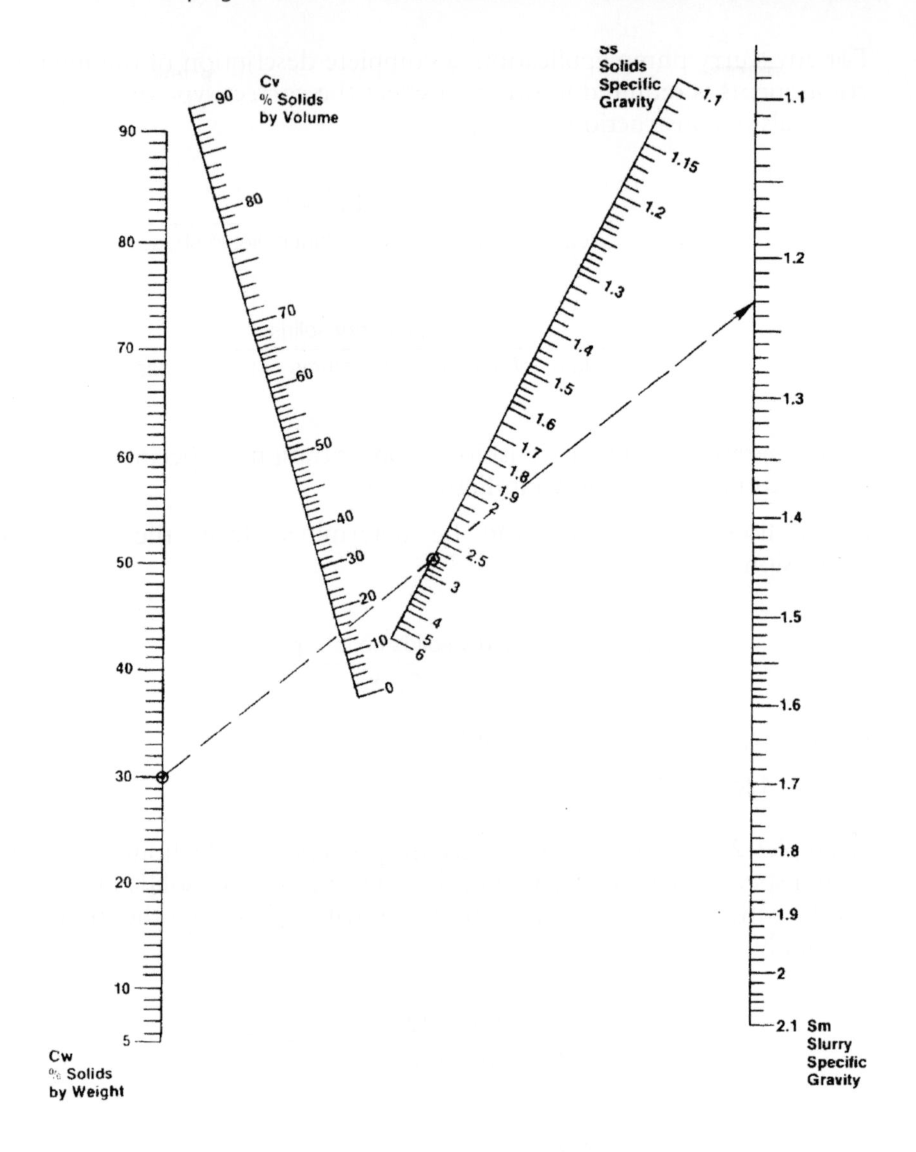

Figure 8.6: Nomograph of relationship of concentration to specific gravity (Reproduced with permission of The Hydraulic Institute)

Wear increases rapidly when the particle size increases. The life of the pump parts can be extended by choosing the correct materials of construction.

Sharp angular particles cause about twice the wear of rounded particles.

Austinetic manganese steel is used when pumping large dense solids where the impact is high.

Hard irons are used to resist erosion and, to a lesser extent, impact wear.

Elastomeric materials are used when pumping concentrations of fine material but total head is usually restricted to about 100 feet per stage.

Castable ceramic materials have excellent resistance to cutting erosion but impeller tip velocities are usually restricted to 100 ft./second.

The last 2 sections are reproduced with permission of the Hydraulic Institute.

8.2 Paper stock

Paper Stock consists of cellulose fibers, up to about 1/4 inch long, suspended in water. Once the stock has been washed or screened to remove unwanted chemicals or impurities, it is either beaten or refined to enhance the sheet properties. Various additives, such as starch, alum, size or clay fillers, are then introduced to create the required characteristics of the paper product being produced.

When pumping Paper Stock, a number of difficulties are experienced, all of which are attributable to the nature of the basic product. If these difficulties were encountered individually, they could be handled quite easily. Collectively, they present a bigger challenge.

The simple presence of the fibers necessitates a pump capable of handling small particles that, in high densities, may have a tendency to clog the impeller. While lighter duty services will use the conventional chemical process pump, higher flows and stock consistencies are handled by a Stock Pump (Figure 8.7). The latter is usually a much more robust design than an equivalent ANSI model.

While the physical size of the individual fibers is not a problem, they do have a tendency to float in water. This requires constant agitation of the stock to minimize the possibility of stratification. Unfortunately, such agitation can introduce air which, in addition to being detrimental to the stock, can cause considerable pumping difficulties.

Paper stock acquires an affinity for water as it is beaten or refined. The retention of water by the stock increases the friction factor which is not usually significant when the piping velocities are maintained at their normal rates. However, at higher velocities, a heavily beaten stock with a low freeness value becomes slippery and difficult to pump.

The consistency of a pulp and water suspension is the percent by weight of pulp in the mixture. Bone Dry (B.D.) consistency is the amount of pulp left in a sample after drying in an oven at 212°F or above. Air Dry

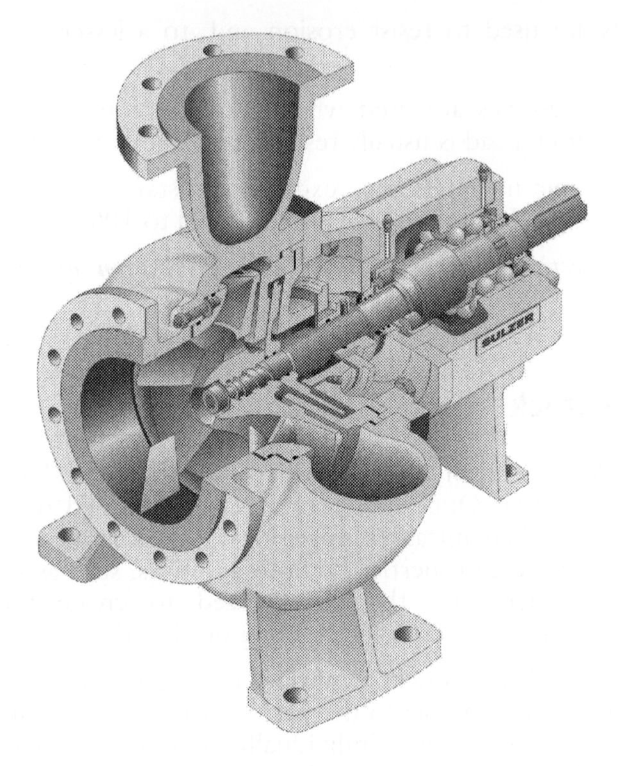

Figure 8.7: Paper stock pump (Reproduced with permission of Sulzer Pumps)

(A.D.) consistency is an arbitrary convention used by papermakers, and is the amount of pulp left in a sample after drying in atmosphere. Air Dry stock contains 10% more moisture than Bone Dry stock, i.e., 6% B.D. is 6.6% A.D. Air Dry consistency is the term most commonly used, and it will be used throughout this discussion unless otherwise noted.

The consistency can range between 0.1% up to about 20%, but only starts to become a pumping problem when it reaches the 3% level. Up to that point, pump performance can be considered to be like water. As the consistency increases, pump performance will decrease rapidly. Between 3% and 6% consistency, correction factors are traditionally used for the performance of a centrifugal pump, as discussed in Chapter 8.2.5.

8.2.1 Air in stock

Entrained air is detrimental to good operation of any centrifugal pump, and can result in reduced capacity, increased erosion and shaft breakage. Obviously every effort must be made to prevent the over-entrainment of air throughout the process.

8.2.2 Excessive discharge throttling

While it is realized that excess capacity is normally required over the paper machine output in tons per day, the 'over-selection' of pumps on the basis of capacity and head usually results in the necessity of throttling the pump at the valve in the discharge line. Since the valve is frequently located adjacent to the pump, the restriction of the valve and the high velocity within the valve will result in some dehydration and cause vibration due to slugs of stock. Vibration at the valve due to throttling is transmitted to the pump and may reduce the normal life of the pump rotating element.

Centrifugal pumps operating at greatly reduced capacity have more sever loading internally due to hydraulic radial thrust as discussed in Chapter 2.5. Hence pumps selected too greatly oversize in both capacity and head have the combination of the vibration due to throttling plus the greater internal radial load acting to reduce the life of the rotating element. As a general rule, stock pumps should not be operated for extended periods at less than one quarter of their capacity at maximum efficiency. When excessive throttling is required, one of the two methods below should be employed.

- Revise capacity requirements and check the static and friction head required for the capacity desired. Reduce the impeller diameter to meet the maximum operating conditions. This will also result in considerable power savings.
- Install a by-pass line upstream from the discharge valve back to the suction chest below the minimum chest level, if possible, and at a point opposite the chest opening to the pump suction. This by-pass line should include a valve for flow reduction. This method is suggested where mill production includes great variation in weight of sheet.

8.2.3 Filters and additives

The presence of fillers and chemical additives such as clay, size and caustics can materially increase the ability of paper stock to remain in suspension. However, overdosing with additives such as alum, may cause gas formation on the stock fibers resulting in interruption of pumping.

The last 3 sections are reproduced with permission of Goulds Pumps, ITT Industries

8.2.4 Paper stock pumps

Standard centrifugal pumps have difficulty handling stock over 6% and consequently special designs are necessary, and are currently referred to as Medium Density and High Density pumps.

8.2.4.1 Medium density stock pumps

Medium density applications can be handled successfully by a centrifugal pump. However, this capability is greatly dependent on the fluidity of the stock and the ability of the system to deliver it freely to the impeller eye.

Special stock pump designs have been produced and are reputed to be able to deliver the same head and flow for stock as for water at consistencies of up to 5–7%. As the power consumption and efficiency remain almost unchanged, this would result in dramatic reductions in horsepower requirements.

Other centrifugal pumps such as that shown in Figure 8.8, are being used on increasingly higher consistencies using a special rotor device extending into the suction line. This device fluidizes the stock and allows it to be handled by the centrifugal impeller.

The piping arrangement on the inlet side of the pump is especially important. Stock must be delivered freely to the pump, and with good cross-sectional distribution, to prevent slugs of stock from entering the impeller eye. The pipe should be a minimum of one to two pipe sizes larger than the pump inlet and should be straight with no high spots where air pockets can collect.

Unlike conventional pumps however, the stock pump often requires a short suction line and a short, stepped suction reducer at the pump

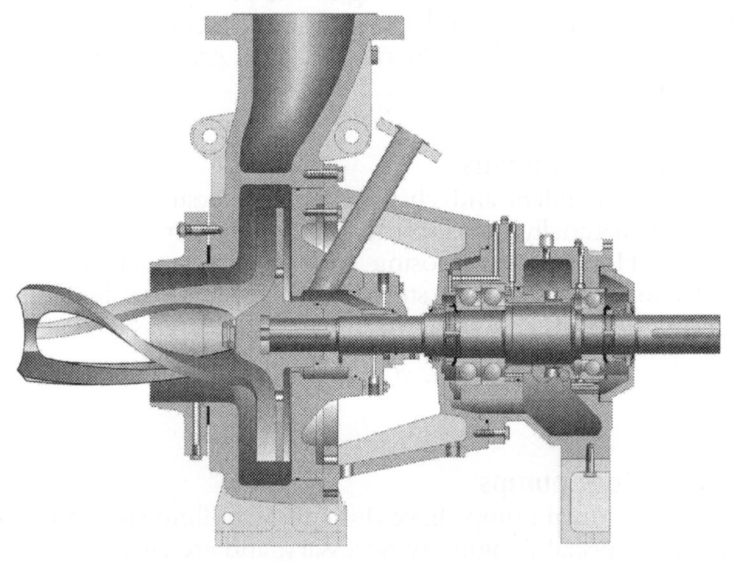

Figure 8.8: Medium density pump (Reproduced with permission of Sulzer Pumps)

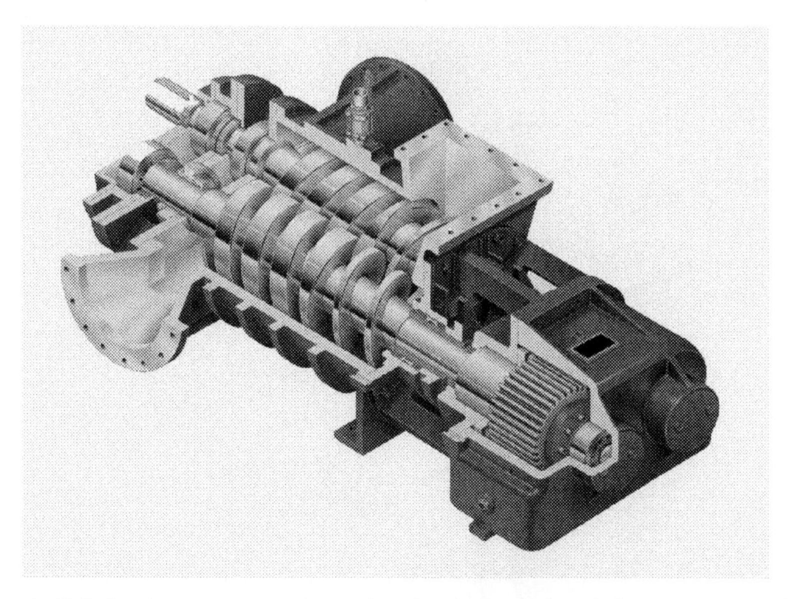

Figure 8.9: High density screw pump (Reproduced with permission of Warren Pumps Inc.)

inlet to distribute the stock in a more uniform manner across the pipe. This prevents the slugging effect as the stock enters the impeller. It also agitates the stock in those cases where it tends to dewater rapidly. For paper stock that has a high tendency to dewater, agitation in the suction source is very important to the operation of the stock pump.

8.2.4.2 High density pumps

The High Density pumps are traditionally a positive displacement design which often uses twin intermeshing pump screws driven by external timing gears to maintain the efficient clearance between the screws. See Figure 8.9.

8.2.4.3 The fan pump

As the final pump in the process, the performance of the Fan Pump is particularly critical. It is a horizontal, double suction centrifugal pump with an axially split casing which provides the optimum of efficiency and reliability. It must also be able to provide a uniform pressure to the headbox, as the quality of the paper can be severely affected by excessive pulsations from the pump. Consequently, these pulsations must be reduced to an absolute minimum.

Pressure pulsations can be initiated by each impeller vane as it passes the cut-water of the volute. The pressure shocks created are about the same as opening and closing a valve in rapid succession and can result in unacceptable ripples in the sheet. Considerable strides have been made

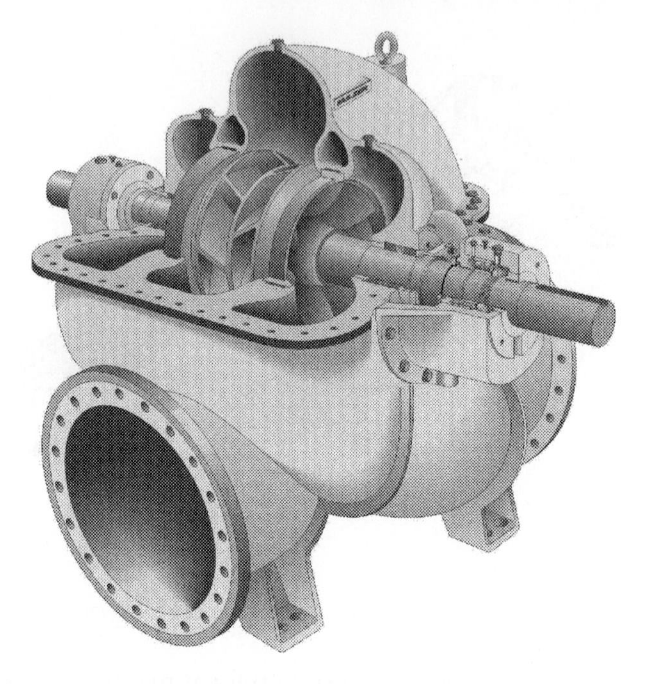

Figure 8.10: Fan pump (Reproduced with permission of Sulzer Pumps)

in reducing these shocks by modifying the vanes to the staggered and angled configuration evident in Figure 8.10.

It is also worthwhile to note that pressure pulsations can also be traced to shaft misalignment between pump and driver, and also to the relative position of the pump operating point to the BEP.

In other words, in addition to all the special considerations identified in this article, a stock pump is still a pump. Consequently it will be dependent on the same conditions of selection, installation, operation and maintenance as all other types of process pumps.

8.2.5 Determination of pump performance

The charts in Figure 8.11 show approximate head-capacity correction factors (Kp) for paper stock suspensions applied to water performance. Correction factors are given at various percentages of maximum efficiency capacity (Qp) for common consistencies.

The correction factors are approximate for normally refined stock of 400–600 SR seconds freeness, with short direct suction lines of ample size. Free or poorly agitated stocks and those with entrained air may require greater corrections. Slow stocks and stocks containing fillers and additives may require less correction.

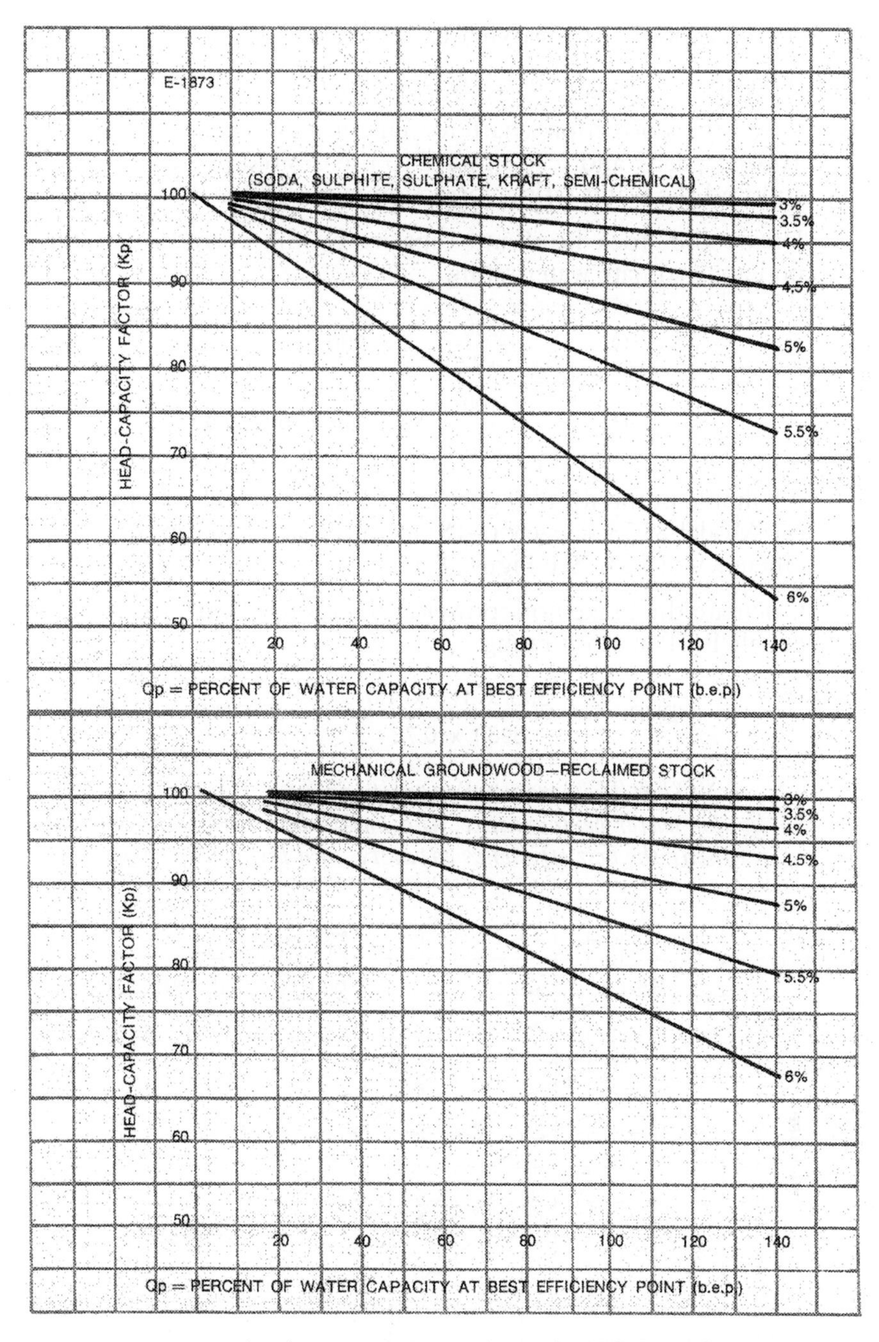

Figure 8.11: Approximate head-capacity correction factors (Reproduced with permission of Goulds Pumps, ITT Industries)

8.2.5.1 Example 1 – Approximate Selection

Select a pump for 600 gpm of 5% A.D. kraft stock at 100 feet total head.

1. Enter the correction chart for chemical stock (Figure 8.11) at 100% capacity and read upward to the desired percent stock (5%) and find the water rating correction factor (Kp = 0.87).

2. Divide the stock head and capacity by the correction factor to find approximate water rating.

$$Q_W = \frac{600}{0.87} = 690 \text{ gpm}$$

$$H_W = \frac{100}{0.87} = 115 \text{ ft Total Head}$$

A suitable pump for 690 gpm at 115 feet Total Head of water is the Model 3175 size 4 × 6–18 operating at 1180 rpm with an efficiency of 73%. See Figure 8.12.

3. Calculate the brake horsepower at the rating, BHP at water rating equals BHP at stock rating.

$$BHP = \frac{690 \times 115 \times 1.0}{3960 \times 0.73} = 27.5 \text{ H.P.}$$

NOTE: The percent of capacity at b.e.p. (Qp) is accurate only for water performance, and should be used only for selection near the best efficiency capacity, and low consistencies.

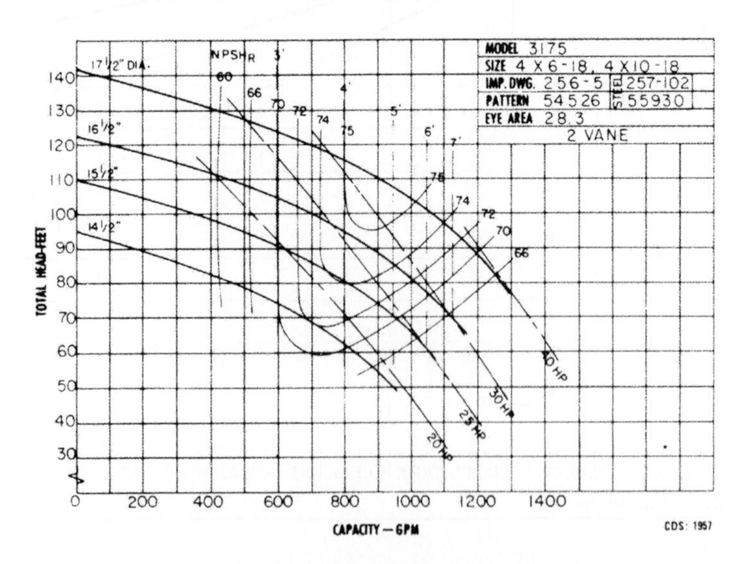

Figure 8.12: Performance curve (Reproduced with permission of Goulds Pumps, ITT Industries)

8.2.5.2 Example 2 – Approximate pump performance on stock when water performance is known.

Find the performance for the Model 3175, 4 × 6–18 at 1180 rpm from curve on Figure 8.12 above, with an impeller diameter suitable for the rating of 600 gpm at 5% Kraft stock at 100 feet Total Head from Example 1.

1. Find the water rating correction factor and approximate water ratings as outlined in Example 1.

   Kp = 0.87
   Qw = 690 gpm
   Hw = 115 ft.

   NOTE: If the approximate water rating for 3.5% stock or over, exceeds the best efficiency capacity, use a larger pump.

2. Refer to Figure 8.12. The capacity at the best efficiency point (b.e.p.) for an impeller diameter slightly smaller than required for the water rating, is 880 gpm. Multiply 880 gpm times the correction factor to find the pump capacity at b.e.p. on stock.

$$880 \times 0.87 = 766 \text{ gpm}$$

3. Find the ratio of the stock capacity at the rating (600 gpm) to the stock capacity at b.e.p. (766).

$$Qp = \frac{600}{766} \times 8 = 78\%$$

4. For capacities not at the b.e.p., the percent consistency is subtracted from Qp found in Step 3 above.

$$Qp = 78 - 5 = 73\%$$

5. Find the correction factor for 73% capacity Qp from the chart, Kp = 0.905

   Divide the rated stock Head and Capacity by the new correction factor to estimate the final water rating.

$$Qw = \frac{600}{0.905} = 663 \text{ gpm}$$

$$Hw = \frac{100}{0.905} = 110 \text{ ft. Total Head}$$

WATER				
Qp	GPM	Total Head ft.	Eff.	BHP
40	348	121	53	20.1
60	522	115.6	66	23.1
80	696	108.6	73	26.1
100	870	99.7	75	29.2
120	1043	87	72.5	31.7

Figure 8.13: Water table (Reproduced with permission of Goulds Pumps, ITT Industries)

Select an impeller diameter for this water rating.

Diameter = 16.875 inches

NOTE: The impeller diameter selected must be approximately equal to the diameter used in Step 2 above. If the diameters vary appreciably, repeat Steps 2, 3 & 4.

6. From Figure 8.12 performance curve, tabulate the water head, the efficiency, the BHP at 40, 60, 80, 100 and 120 percent of capacity at b.e.p. for the diameter impeller required for the final rating in Step 5 above.
7. Tabulate correction factors at the percent capacities above from the correction chart.

STOCK				
Kp	GPM	Total Head ft.	BHP	Eff.
0.948	330	115.0	20.1	47.7
0.922	481	106.5	23.1	56.0
0.896	624	97.3	26.1	58.7
0.870	757	86.7	29.2	56.8
0.844	881	73.4	31.7	51.5

Figure 8.14: Stock table (Reproduced with permission of Goulds Pumps, ITT Industries)

8. Calculate stock head-capacities by multiplying water performances by the correction factors. Calculate stock efficiencies using equivalent water BHP values.

$$\%Eff._{(stock)} = \frac{GPM_{(stock)} \times Total\ Head_{(stock)} \times SP.\ Gr.}{3960 \times BHP_{(stock)}}$$

NOTE: BHP (Stock) = BHP (Water)

9. Plot the curve for water and the approximate stock performance as in Figure 8.15.

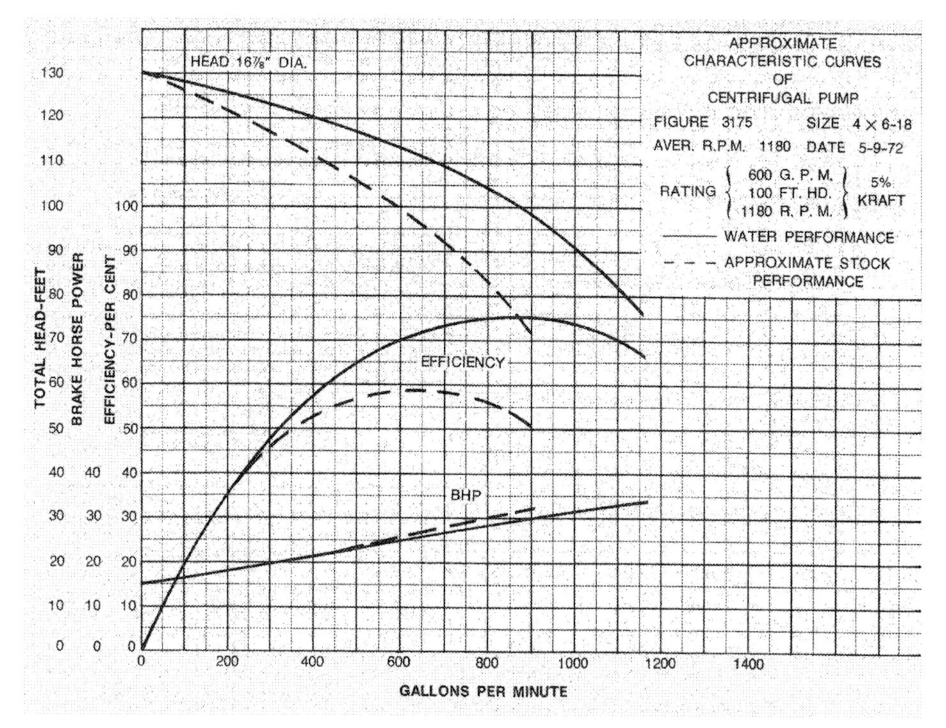

Figure 8.15: Approximate characteristic curves of centrifugal pump (Reproduced with permission of Goulds Pumps, ITT Industries)

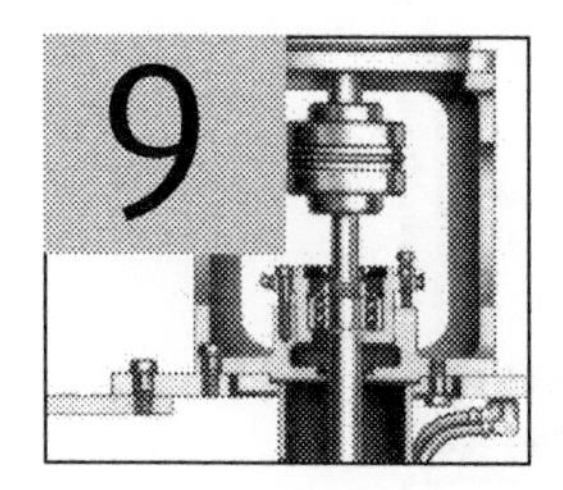

Special pumps

9.1 Sump pumps

When a sump or wet well has to be pumped out frequently, or have the level controlled on a regular basis, a variety of pump styles are available to do the job. The actual pump selected will depend on a number of factors including the past experience of the engineer, the size and type of operation, and the budget. As we have no control over most of these, let's consider the various pump styles that are in popular use in various industries.

9.1.1 Vertical sump pump

The sump pump is suspended from a sole plate mounted across the top of the sump, with the wet end submerged below the level of the liquid in the sump. The nature of this design lends itself to these applications where the sump should be covered for safety or other reasons, as the sump cover can be designed to act as a base for the pump soleplate.

This style frequently uses the same impeller and casing as end suction pumps. The back cover and stuffing box are replaced by a cover fitted with a steady bushing while the shaft is changed to the long column shaft that runs up inside the column assembly. That assembly accommodates the sleeve bushings, and both the column itself, and lineshaft bushings are usually spaced at 1ft. and 3ft. intervals.

Depending on the nature of the liquid being pumped, the lineshaft bushing can be lubricated by the pumpage, or by an injection of grease or oil. The outside lubricants are delivered to the bushings through tubes which are connected to the pump soleplate for ease of lubrication.

In this pump design, the liquid is not pumped through the column assembly. The discharge nozzle on the pump casing is horizontal and discharges into an elbow that directs the liquid through a vertical pipe to an above grade connection at the soleplate. This dual connection of the discharge pipe and column to the sole plate provides additional rigidity for the pump to help withstand the effect of the radial loading in the pump casing.

As the operation of the pump usually depends on the level of liquid in the sump, a variety of float switches can be used for automatic on/off control. In the event that the level in the sump is likely to draw down below the level of the impeller, a tail pipe can be bolted to the suction flange that will allow the pump to operate under these conditions.

A drainage type of application constitutes the vast majority of services for which these pumps are purchased. In such conditions they should not be considered or treated as a heavy duty pump. However, a number of manufacturers have dramatically upgraded the design for operation in chemical process services where the liquid being pumped is so aggressive that safety standards prohibit any connections in the storage tank below the level of the liquid.

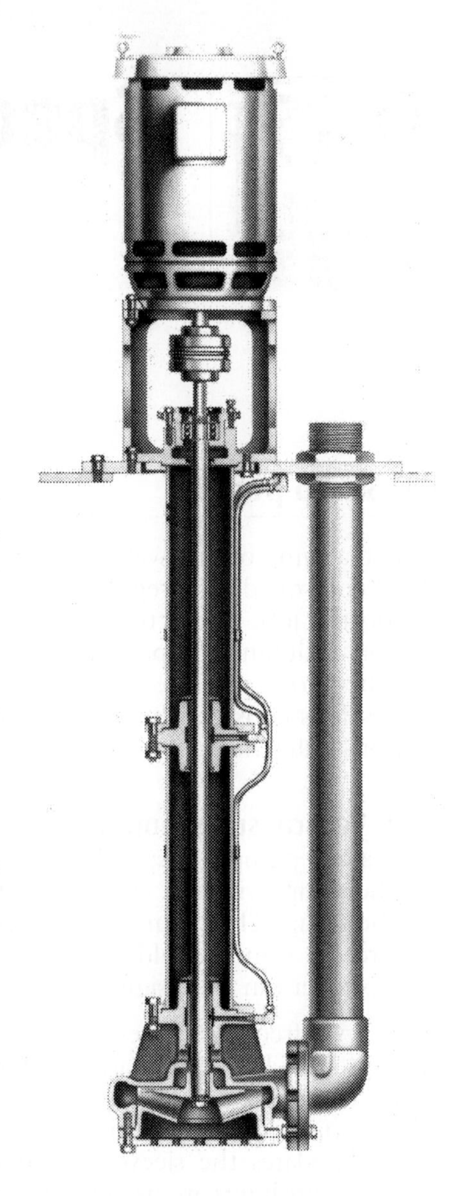

Figure 9.1: A typical sump pump (Reproduced with permission of Goulds Pumps, ITT Industries)

9.1.2 Submersible pump

Another popular option is the submersible pump that is essentially a close coupled unit where the impeller is mounted directly on the motor shaft, and the casing is bolted to the motor. As a consequence of this design, the motor shaft must be able to resist all the impeller radial and axial forces, and the motor bearings must carry the pump impeller loads in addition to normal motor electromagnetic and torsional loads.

The submersible pump incorporates a motor that is designed so that the entire unit can operate while immersed in the pumped liquid. The compact design and maneuverability of this pump results in very flexible usage in a variety of applications, but mostly in drainage types of service.

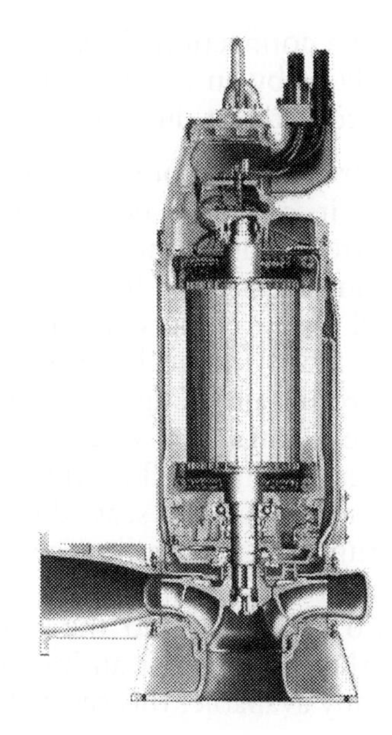

Figure 9.2: Typical submersible pump (Reproduced with permission of ITT Flygt)

In the water and waste treatment industry, these pumps are occasionally secured in a fixed position at the bottom of the wet well and attached to the fixed discharge piping. More frequently they are used in conjunction with a set of slide rails and a special discharge flange connection that allows them to be drawn up from the bottom of the wet well for service. Sometimes they are used as portable units and simply lowered into place as required. As with other submerged suction pumps, a variety of float switches are available for automatic on/off control.

If these pump types are used in a critical application, it is worthwhile to remember that all the usual cautionary signs of pump failure such as vibration, noise and overheating, will go undetected because the pumps are operating under water. Under these conditions, it is recommended that some kind of alarms be included to indicate such problems.

9.1.3 Priming the pump

Pump styles which do not operate with a submerged suction can also be utilized. These pumps will be located above the sump, with only a suction line being lowered into the sump, and with the liquid source below the level of the pump. This is known as operating on a suction lift and, under these conditions, it is important to be able to 'Prime' the pump to ensure that it starts to pump.

Priming the pump requires removing all the air from the suction line and the pump, and filling them with water. This is usually achieved through a connection close to the discharge nozzle of the pump, either manually or by an automated system. At that point the pump can then be started without difficulty and will operate under normal conditions.

While priming the pump is always essential on installation and initial startup, it can also be necessary when the pump is operating on intermittent duty. Under these conditions the pump will occasionally stop, thus leaving the liquid in the suction line to the mercy of gravity that will tend to empty that line. To prevent such an occurrence, a foot valve is fitted to the bottom of the suction line in the sump. These valves are usually designed with a flap or ball positioned in such a way that, the normal flow direction of the liquid will hold the valve open. However, when the flow through that valve reverses, the flap or ball will close against the seat, thus holding the liquid in the suction line. Unfortunately, if a small piece of debris is caught between the seat and the flap or ball, the valve will stick open and the suction line will drain out into the sump and the pump loses its prime. For occasional use where the pump is not an important part of the process, and operators are readily available, this repetitive priming may be quite acceptable.

9.1.4 Air operated double diaphragm pump

A variety of positive displacement pumps have the ability to prime themselves, and information on them can be seen in Chapter 9.4.

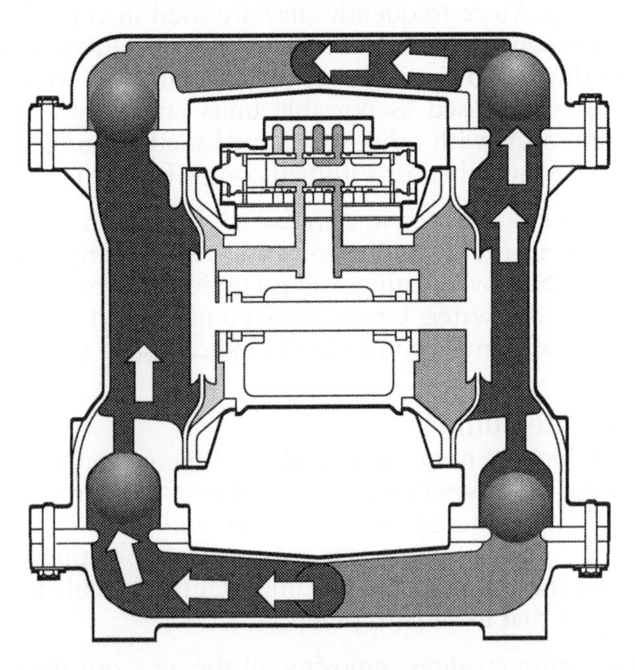

Figure 9.3: Double diaphragm pump (Reproduced with permission of Warren Rupp Inc., a Unit of IDEX Corp.)

However, the one that is most commonly used in drainage sump applications is the air operated double diaphragm pump.

This pump style is essentially two pumps in one where one is on the suction cycle while the other is on the discharge cycle. Compressed air is delivered through the air shuttle valve and is directed alternately from one diaphragm to the other. As it pressurizes one diaphragm, it simultaneously exhausts the air from the other.

When a diaphragm is pressurized it moves inwards towards the pumping cavity and raises the pressure in that cavity. This closes the suction valve against the lower suction line pressure, and opens the discharge valve to force the liquid out of the pump. As the diaphragm moves out from the pump cavity, the pressure is lowered, thus allowing the suction valve to open and the discharge valve to close. This movement creates a low pressure area in the pump cavity as the air is forced out, thus allowing the liquid to move up the suction line into the pump.

9.1.5 Horizontal centrifugal pump

A horizontal centrifugal pump can be used on a suction lift application, but only with some assistance to ensure that it is always primed prior to startup.

The Air Ejector System shown in Figure 9.4, uses available compressed air to vacate the entrained air from the suction line and pump prior to startup. When the air ejector starts to emit a steady stream of liquid that is the indication that all the air is evacuated and the unit is filled with liquid, and ready to start.

Another method to maintain prime uses a Priming Tank at the suction side of the pump as shown in Figure 9.5. This priming tank must be

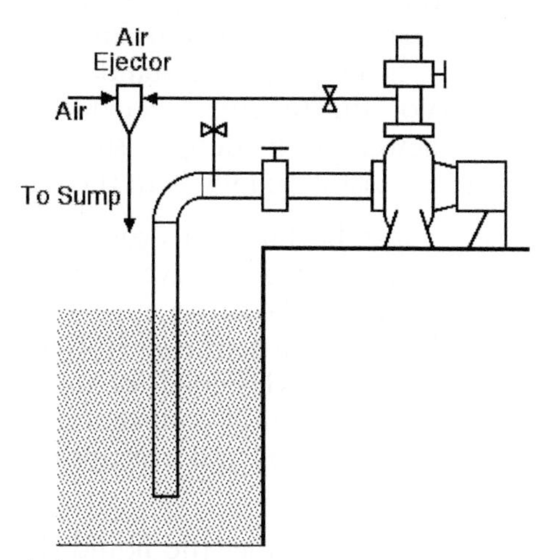

Figure 9.4: Typical air ejector priming system

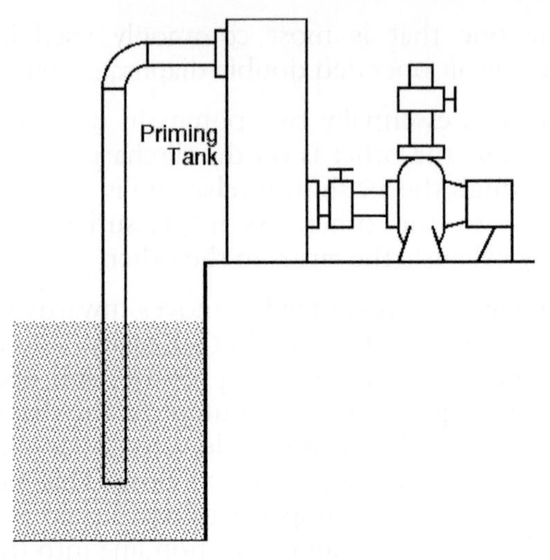

Figure 9.5: Priming tank arrangement

sized so that it contains 3 times the volume of liquid contained in the suction line. When the pump is started up while the suction line is empty, the amount of liquid in the priming tank will be displaced by the pump and creates a low pressure area in the tank. This allows the liquid in the sump to be moved up through the suction line and enter the priming tank.

The priming tank must also be capable of supplying sufficient NPSH to the pump as it essentially replaces the sump as the suction source.

9.1.6 Self-priming pump

One of the more popular methods of dealing with this problem however, is with the use of a self-priming pump that is capable of freeing itself of entrained air and resuming normal pumping without any supervision. It is essentially a standard end suction centrifugal pump with two major attachments; a Suction Reservoir and a Priming Chamber. In the newer pump designs such as shown in Figure 9.6, these attachments are cast integrally with the pump casing.

The pump casing must be filled in order for the pump to prime at the initial startup. From that point, every time the pump shuts downs, the reservoir retains enough liquid to automatically prime the pump at the next startup, if it should lose the suction leg or become air bound. Most modern pump casings, designed for solids handling service, have a secondary internal recirculation connection that allows the liquid to drain back into the volute scroll and exit tips of the impeller during the priming cycle.

When the pump loses it's prime, the liquid drains out of the suction

Figure 9.6: Self priming pump (Reproduced with permission of Gorman-Rupp Pump Company)

pipe, but water is retained in the suction reservoir as a result of the elevated suction inlet. When the pump is restarted, the mixture of air and liquid from the suction reservoir is pumped into the priming chamber. At that point the velocities decrease, allowing the air and liquid to separate.

The air bubbles up and is released through the discharge nozzle where it will escape freely if the system is open to atmospheric conditions. In the event of a pressurized operating system however, an air release mechanism must be provided at the pump discharge.

At the same time, the heavier liquid drops to the lower part of the priming chamber where it drains back through the internal recirculation channel to the tips of the impeller. At that point, air inside the impeller mixes with the returned water and causes a pressure reduction in the eye of the impeller as the air/water mixture is repeatedly expelled from the impeller. This continues throughout the priming cycle until all the air is removed and the pump and suction line are filled with liquid. At that point, a normal pumping cycle will be re-established.

Many self-priming pumps also use a check valve at the inlet to the suction reservoir to prevent the loss of the suction leg and repeated priming on intermittent service.

Self-priming pumps are available in a number of designs to accommodate the size and type of solids entrained in the pumpage.

9.2 Vertical turbine pumps

Vertical Turbine Pumps have come a long way since their humble beginnings in the irrigation field, and are now used in a wide variety of applications. While they are still utilized to raise water from underground aquifers, ponds, lakes, rivers and oceans, they now supply such industries as Water Treatment Facilities, Pulp and Paper Mills and Power Generating Stations. They are also an integral part of numerous Fire Protection Systems, as well as many of the processes in the Petrochemical Industries.

These vertical diffuser pumps come in a variety of configurations. The lineshaft driven Deepwell Turbine Pump shown in Figure 9.7 is used in a wide variety of applications and is usually built to AWWA standards. The Submersible motor driven pump is most frequently used to raise well water from underground aquifers through drilled wells in rural communities. Materials of construction of these pumps are mainly steel or cast iron bowls and discharge heads, and bronze impellers.

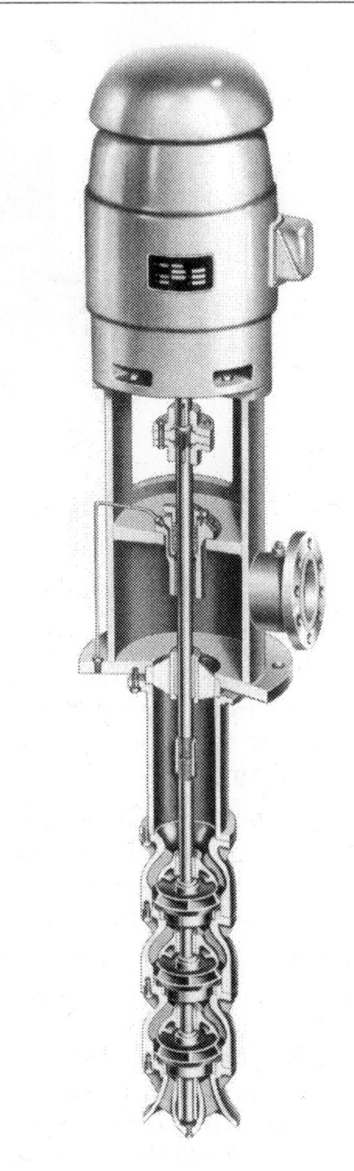

Figure 9.7: Deepwell turbine pump (Reproduced with permission of Goulds Pumps, ITT Industries)

The larger Service Water Pumps are primarily used on municipal and industrial water intake services. In addition to the cast iron bowls and bronze impellers, they are frequently used with stainless steels or nickel aluminum bronze. The columns and discharge heads are usually supplied in steel, and can be designed with the discharge nozzle located in the column below the mounting plate when required.

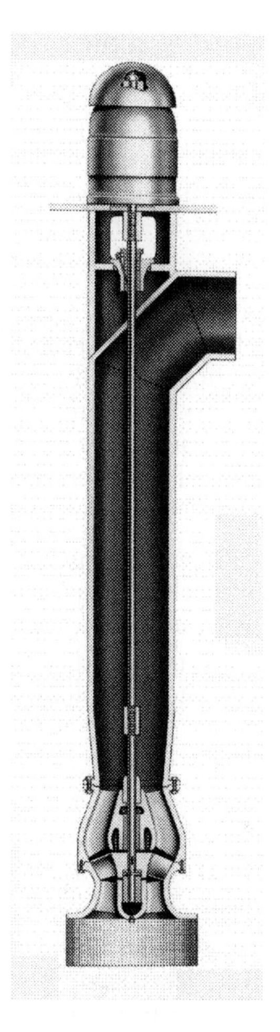

Figure 9.8: Service water pump (Reproduced with permission of Goulds Pumps, ITT Industries)

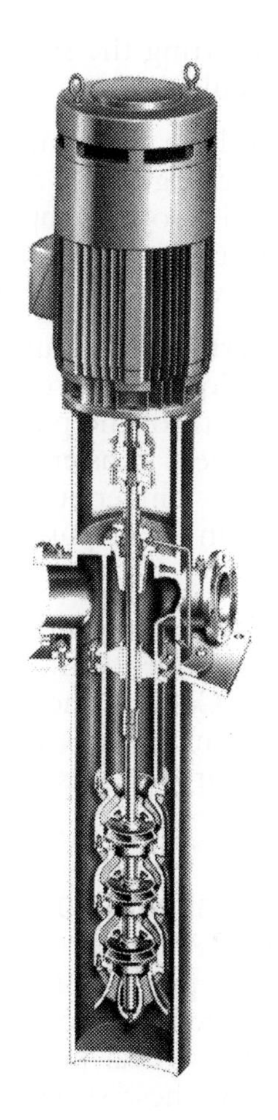

Figure 9.9: Vertical can pump (Reproduced with permission of Goulds Pumps, ITT Industries)

Most of these pumps have motor drivers with thrust bearings in the motor. A hollow shaft motor is frequently used, and this allows the pump lineshaft to pass through the motor and be secured by the thrust bearing at the top. Others use a solid shaft motor with the coupling located in the discharge head assembly above the stuffing box.

The Vertical Can pump shown in Figure 9.9 incorporates the Column and Bowl Assembly within a pressure containing suction vessel that is constructed in accordance with ASME Section VIII. These pumps are widely used in the Petrochemical industry as pipeline or booster pumps where the NPSH available is very low. This design increases the NPSH available by lowering the first stage impeller into a longer suction can, thus permitting the suction source to be emptied to a theoretical zero NPSH available.

Many of these pumps use a solid shaft motor with the coupling located in the discharge head assembly above the stuffing box. Frequently the pump will also be supplied with the thrust bearing in the same area, although this increases the total height of the installation which could aggravate vibration problems.

These vertical can pumps can be modified to handle temperature sensitive liquids. For example, sulfur can be handled by jacketing the entire length of the pump and using steam or hot oil in the jacket. In addition, heat transfer fluid pumps can be built to withstand the forces caused by temperature variations and pipe strain. The bowls and impellers in these pumps are usually made out of 300 series stainless steel.

The typical vertical turbine pump is made to order and consists of three major components; the Bowl Assembly which uses a variety of standard parts, the Column and the Discharge Head. In view of this, a high level of communication is essential between the end user and supplier to ensure that the best equipment is selected and supplied for the application.

9.2.1 Vertical pump bowl assembly

The Bowl Assembly consists of the impeller and the bowl. As in the horizontal pumps, the impellers can cover the entire spectrum of specific speed as discussed in Chapter 1.4.1. Most commonly used in the larger vertical pumps however, are the mixed flow or axial flow designs. As these styles usually develop relatively low heads, the use of multistage pumps is common where higher pressures are required. These impellers are all keyed to a common shaft or secured by means of tapered colletts.

The pump bowl contains the axial diffuser vanes to direct the flow of liquid from one impeller to the next or into the pump column. They also convert the velocity energy as it is leaving one impeller, to the pressure energy at the inlet of the next impeller. These bowls can be either flanged or screwed together, depending on the particular pump design. The bowls also contain the sleeve bearings which act predominantly as guide bushings only, as the axial diffuser design minimizes radial thrust from the impellers.

In the larger pumps the inlet to the bowl assembly is fitted with a suction bell which provides a smooth flow transition into the first stage from an open pit type of application. This also applies to those pumps used as main intake pumps from a well screened sump. Other designs include a suction casing fitted with a suction strainer or with a longer

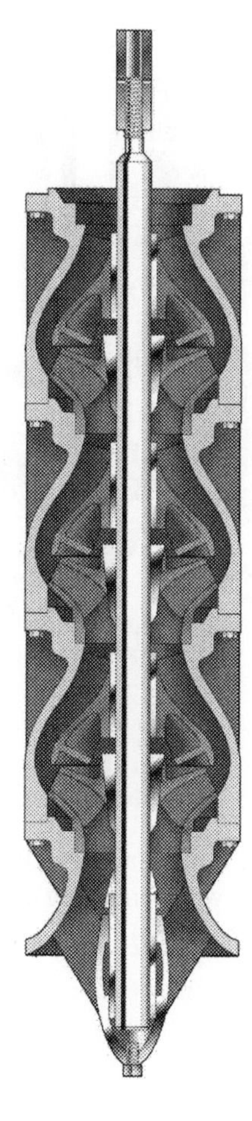

Figure 9.10: Typical bowl assembly (Reproduced with permission of Goulds Pumps, ITT Industries)

suction pipe that may be needed in well pumping when the water level in the well might drop below the normal inlet.

These pumps that utilize suction cans in various industry processes tend not to use the suction bell as it would take up valuable radial space in the can. When the process involves a fairly clean liquid, the suction screen will not be used.

9.2.2 Vertical pump column

The length of the column establishes the depth of the pump and comes in various sections that are usually standardized to 5 feet and 10 feet lengths. It is available in either screwed sections or flanged connections. The column connects the bowl assembly to the discharge head and encloses the line shaft that is connected to, and drives the pump shaft. The line shaft sections are connected by threaded or sleeve type couplings.

In clean applications, the shafting is left exposed to the pumpage moving through the column and the bearing bushings are supported by bearing retainers, with sleeves protecting the shaft in the way of the guide bushings. An enclosing tube can be fitted to isolate the shafting from the pumpage and provide a means by which to lubricate the bearings from an external source with clean water, grease or oil. Stabilizers are included to support the tube in the column and minimize the possibility of vibration problems.

The most common materials used in lineshaft bearings on water service are cutless rubber and bronze. While both of these are widely used in cold water applications, the cutless rubber is better suited to water which may include some sand. Carbon bearings are also used in more corrosive applications and when pumping chemicals with low lubricating properties.

9.2.3 Pump discharge head

The discharge head can be either a fabricated part of a casting that provides a base from which the pump is suspended and on which the driver is mounted. It also directs the flow from the vertical column to a horizontal flanged discharge nozzle that will be connected to the system. The discharge head is fitted with a stuffing box or seal chamber through which the line shaft passes to connect with the driver. Depending on the application, either packing or mechanical seals can be used. When an enclosing tube is used, a tension nut is fitted to keep the tube under tension and to hold it straight.

Figure 9.11: Vertical pump discharge head (Reproduced with permission of Goulds Pumps, ITT Industries)

9.2.4 Vertical pump performance

The performance of the vertical turbine pumps using the mixed flow or axial flow impellers can develop some significant operational differences, particularly in the larger sizes.

In the performance curve shown for a typical mixed flow impeller (Figure 9.12) it can be noted that the shutoff head may be more than twice the head at the best efficiency point. An axial flow impeller will produce an even higher difference at the shutoff condition. As the power draw is also at its maximum at the shutoff condition, an automatic by-pass will be needed in the event the discharge is clogged or restricted by valving. Otherwise the driver will be overloaded.

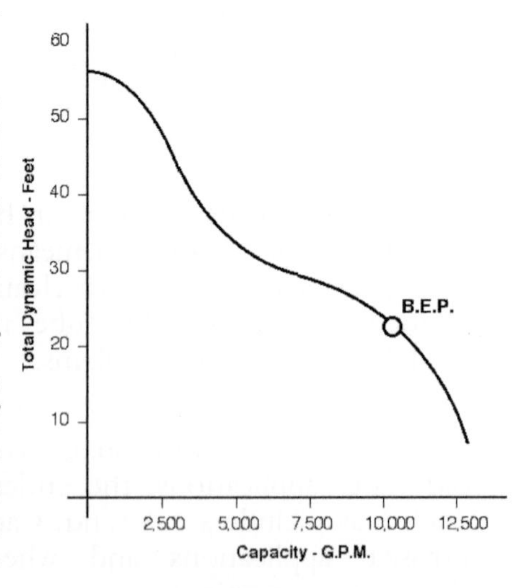

Figure 9.12: Vertical turbine pump curve

Although the NPSH required rises quickly with increased flow rates, this is rarely a problem when pumping from an open sump. However, the matter of submergence must always be considered as it is independent of the NPSHR. The submergence is the static elevation difference between the free surface of the liquid and the centerline of the first stage impeller. It is interesting to note that the submergence required by all vertical pumps can be represented on the pump curve by a horizontal straight line – but only up to a point. Beyond that point the Submergence required increases dramatically, causing a vortexing problem where air will be drawn into the impeller.

9.2.5 Vertical pump sump requirements

Vertical pump performance is also dependent on sump design as this determines the flow of liquid to the pump. Unfortunately this is much less precise than the inlet piping of a horizontal pump and frequently results in performance problems. The layout of the sump and the relative positioning of the pump(s) within that sump are of such importance to the pump reliability as to warrant model testing prior to the design and installation of the sump. The objective is to provide a sump in which there is minimal turbulence and no negative influence from the incoming lines.

Figure 9.13 shows a possible installation with some very general guidelines that would apply to the particular layout shown. Some of the key elements include the velocity of the approaching water which should usually be about 1.0 ft/sec., and the location of each pump to

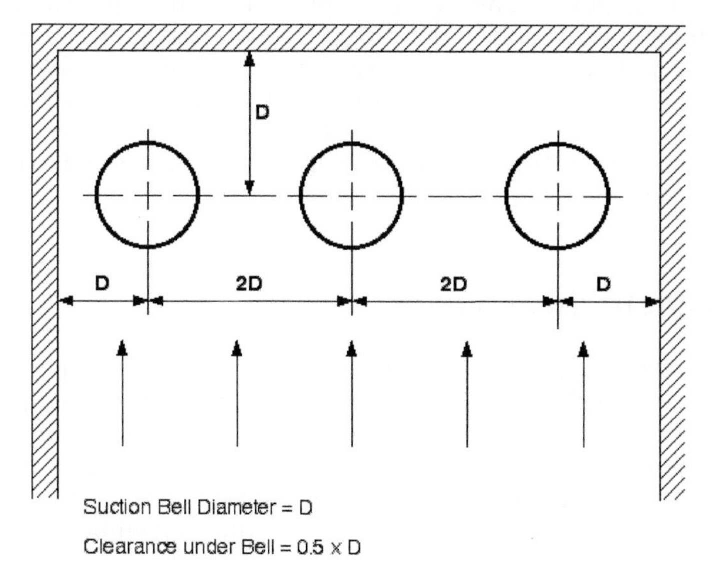

Figure 9.13: Vertical pump sump layout

the other(s) and to the side and back walls of the sump. These distances are also related to the entrance velocity of the pump which in turn can be a function of the suction bell diameter. The clearance of the bottom of the pump to the sump floor is generally considered to be about half of the suction bell diameter of the pump. It should be noted that bell diameters for the same size and type of pump may vary from one manufacturer to another.

It should also be noted that different arrangements of pumps and accompanying baffles are used in sump designs. Many installations are now using dividing walls between each pump to contribute to the smooth flow of liquid to the suction bell of each pump, as this is necessary for optimum operation. It is particularly important to avoid any cascading effect from incoming liquids that might create air entrainment or vortexing that will detrimentally affect the pump performance.

9.2.6 Vertical pump design considerations

Owing to the wide diversity of applications, there are numerous design options available on a Vertical Turbine Pump. While this provides a high degree of flexibility in the selection process it should also raise a warning flag to those who are constrained by budgetary limitations. Under these conditions it is particularly important to ensure that the pump is accurately specified to reflect the need of the application.

One of the most vital considerations must be the assurance and reliability of part straightness, concentricity and parallelism throughout the life of the pump. This particular problem stems from the fact that the shafting must be concentric with the column assembly and the bowl assembly, and that these two assemblies must also be fully parallel with each other and with their own individual parts. Although that involves a lot of accuracy needed during both installation and maintenance of a large pump, it will be well rewarded with high reliability and long life.

9.3 Magnetic drive pumps

Magnetic Drive Pumps were originally designed to pump toxic and other dangerous fluids without the use of mechanically sealed pumping units. The pumpage is retained inside the pump casing and a containment shell, while the impeller shaft is supported on sleeve bearings which are lubricated by that pumpage. The impeller shaft is driven by a magnetic field passing through the containment shell from the driver shaft.

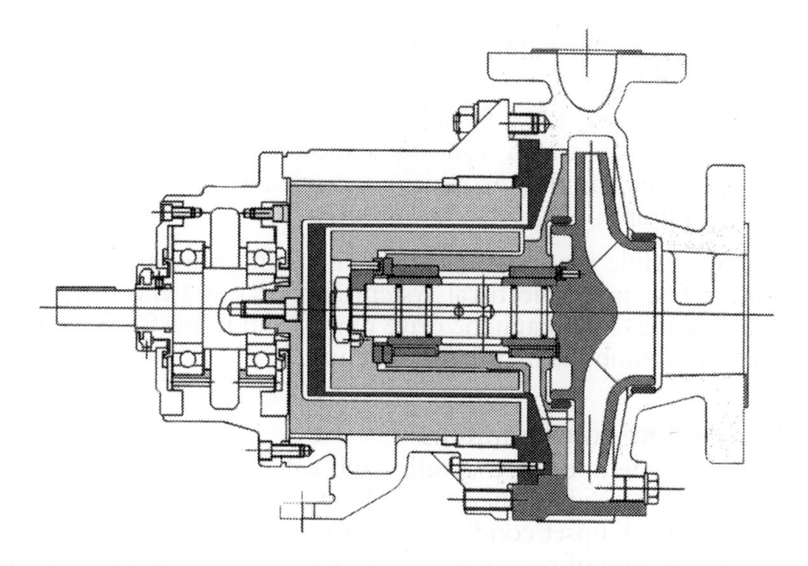

Figure 9.14: Magnetic drive pump (Reproduced with permission of Magnatex Pumps Inc.)

The containment shell may be constructed out of metal or nonmetallic materials. However, when a metal containment shell is used, adequate removal of the heat generated by eddy currents is an important consideration. As this heat generation is proportional to the square of the speed of the pump, it becomes even more vital when the liquid being pumped is sensitive to temperature changes.

9.3.1 Pump bearings

A major difference between magnetic drive pumps and the conventional style is the location and type of bearings. In conventional pump designs, the bearings are usually located well away from the pumped liquid in a well controlled environment. In addition, the operator has a wide choice of appropriate lubricants that can be utilized. With magnetic drive pumps however, the bearings on the impeller shaft are lubricated by the pumpage that may not be an appropriate lubricant. In addition, when the pump runs dry, or operates at very low flows, the lubricant tends to disappear and the bearing will overheat. Even if the pump survives that abuse it may still not last very long if such a condition is followed by a cool liquid entering the pump. Under these conditions the bearings will be subjected to a thermal shock which could cause cracking or total destruction.

It is also important to note that, as the bearings are usually of the sleeve type with slots or grooves to supply the lubricant to the bearing running surface, any solids in the pumpage will be detrimental to the bearing reliability. The material of these bearings is usually either Silicon Carbide or Carbon, and while the Silicon Carbide is a longer wearing material, it is also much more brittle and susceptible to thermal and mechanical shock.

Some designs also provide a flexible mounting for these bearings with the use of O-rings which can be subject to the same chemical compatibility issues as they would in a mechanical seal installed in a conventional pump.

Consequently, while the mechanical seal in a conventional pump tends to act like a fuse in an electrical system and becomes the first failure point, the bearings in a magnetic drive pump tend to perform the same function and become the first point of failure.

9.3.2 Temperature considerations

Magnets are also temperature sensitive and will demagnetize if exposed to temperatures exceeding their upper limit. This provides yet another reason to avoid any upset condition that would cause the generation of heat within this type of pump. Such conditions would include running the pump dry or against a closed discharge valve.

To provide some degree of protection against this problem, the material of the magnets should be selected to be able to handle 25 to 50 Fahrenheit degrees above the expected maximum operating temperature. It should be noted that, of the two most common types of permanent magnets used in magnetic drives, samarium cobalt has a higher temperature rating than neodymium iron boron, but has only 60% of the strength. It is also more expensive.

9.3.3 Decoupling

All magnetic couplings are rated for a maximum torque capability beyond which the magnets no longer operate at the same speed. This is referred to as 'decoupling' and, if the pump operates in this state for very long, the magnets will be permanently demagnetized. Consequently, the magnetic drive pump is particularly vulnerable to any abnormal operating conditions that might result in an excessively high torque demand.

An excessive load demand can be created by a variety of hydraulic conditions or a change in the nature of the liquid. It can also be caused by running the centrifugal pump at a higher capacity. An increase in specific gravity or viscosity of the pumpage will also increase the power draw and the load.

In spite of this, most magnetic drive pumps are normally designed to operate with a safety factor that is frequently less than the safety factor of the mechanical shaft coupling or the electric motor driver, thus leaving the magnetic coupling itself in the most vulnerable position for failure. Consequently, the use of power monitors is recommended for all applications in which magnetic drive pumps are used.

9.3.4 Failure costs

When failures do occur, magnetic drive pumps are usually much more expensive to repair than conventional pump designs with mechanical seals. The parts that frequently need replacement can cost a very high percentage of the entire pump replacement value.

Many end users approach the magnetic drive pump as a cure-all for all pump ills. That is not true. In fact, they have less tolerance for misapplication and process upsets than conventional pumps. However, with an understanding of their limitations and unique advantages, they can provide reliable operation.

9.4 Positive displacement pumps

Positive Displacement (P.D.) pumps operate with a series of working cycles where each cycle encloses a certain volume of fluid and moves it mechanically through the pump into the system. Depending on the type of pump and the liquid being handled, this happens with little influence from the back pressure on the pump.

While the maximum pressure developed is limited only by the mechanical strength of the pump and system and the driving power available, the effect of that pressure can be controlled by a pressure relief or safety valve.

A major advantage of the P.D. pump is that it can deliver consistent capacities because the output is solely dependent on the basic design of the pump and the speed of the driver. This means that, if the liquid is not moving through the system at the required flow rate, it can always be corrected by changing one or both of these factors.

To varying degrees, positive displacement pumps are suitable for handling highly viscous liquids. They are also self-priming and therefore have the ability to handle liquids with a certain volume of entrained air.

All discharge valves installed with any PD pump must be open before that pump is started. This will prevent any fast buildup of pressure that could damage the pump or the system.

9.4.1 The piston pump

One of the earliest and most basic type of PD pump, the Piston Pump uses a plunger or piston to force liquid from the inlet side to the outlet side of the pump.

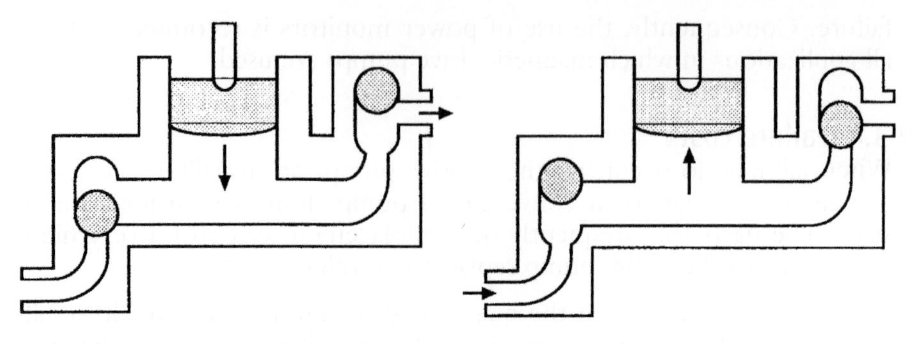

Figure 9.15: Piston pump downstroke

Figure 9.16: Piston pump upstroke

When the piston moves downwards, it raises the pressure in the body which closes the suction valve against the lower suction line pressure, and then opens the discharge valve to force the liquid out of the pump.

When the piston is drawn upwards, it lowers the pressure in the body, which opens the suction valve to admit the liquid to the pump, and allows the higher discharge line pressure to close the discharge valve.

One of the earliest styles of piston pump was the steam-driven piston pump that is still available in both the horizontal and vertical configuration. By arrangement of the valving system, this unit can pump on both the upstroke and downstroke in a manner similar to the air operated double diaphragm pump discussed in Section 9.1.4 and below.

9.4.2 The diaphragm pump

A single diaphragm pump can be similar to the piston pump except that the reciprocating motion causing movement of the liquid through the pump is created by the diaphragm instead of by a plunger. Larger models of this kind of pump are used to pump heavy sludges and debris laden wastes from sumps and catch basins.

Small models of the same basic design are used as chemical metering or proportioning pumps where a very constant and specific amount of liquid is required to be pumped.

A more common type of diaphragm pump is the air-operated double diaphragm pump which uses pressurized air to actuate the diaphragms instead of a mechanical device. This is basically two pumps in one where one is on the suction cycle while the other is on the discharge cycle. The air valves alternately pressurize the inside of one diaphragm chamber and exhaust air from the other one.

This pump does not require additional sealing devices. It can also be operated safely against a closed discharge valve as the air pressure automatically balances out on each side of the pump diaphragms, thus stalling the pump.

Figure 9.17: AODD pump (Reproduced with permission of ProSpec Technologies Inc.)

9.4.3 The gear and lobe pump

The external Rotary Gear Pump is a positive displacement pump where the unmeshing of the gears produces a partial vacuum to draw the liquid into the pump. The liquid carried between the gear teeth and the casing to the opposite side of the pump where the meshing of the gears forces the liquid through the outlet port. The direction of rotation determines which of the nozzles will be the inlet and which will be the outlet. Therefore, by reversing rotation, the pump will be able to pump backwards.

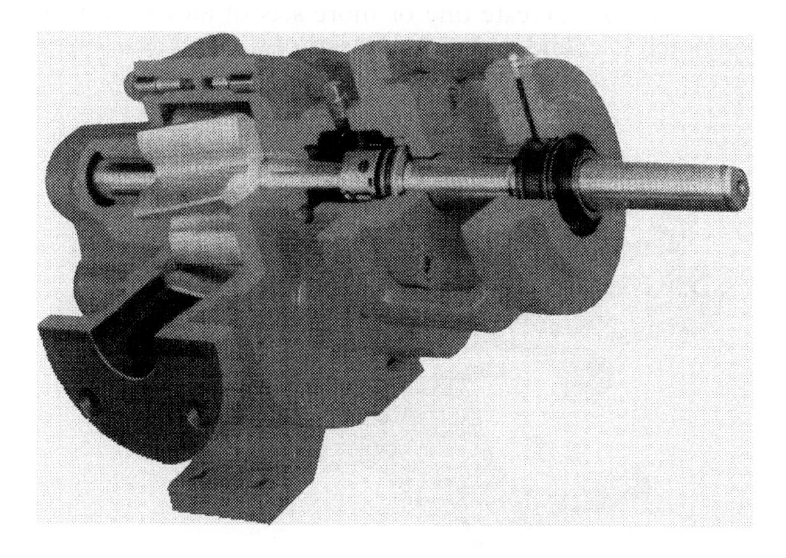

Figure 9.18: External gear pump (Reproduced with permission of Roper Pump Company)

Where one gear is driven by the other, the driven gear usually runs in sleeve type bearings. The bearings and shaft journal are located in the pump casing and flooded by the pumped fluid. Consequently, these bearings and gears are dependent on the lubricating qualities of the pumped fluid.

In other types, the gears have no metallic contact with each other and both rotors are driven by synchronized driving gears separated from the pump chamber. As both shafts pass through the pump casing, two sets of seals are required.

The absence of metallic contact between the surfaces of the rotors and the casing means that the only wear that should occur will be due to friction with the pumped fluid.

A popular modification of the gear pump has some of the gear teeth blended together to form a Lobe Pump that provides a lower shearing action on the pumpage.

The Internal Gear Pump has one rotor with externally cut gears running in the bore, and meshing with, a second internally cut gear.

9.4.4 Screw pumps

Screw pumps are a special type of rotary positive displacement pump in which the flow through the pumping elements is truly axial. The liquid is carried between screw threads on two or more rotors and is displaced axially as the screws rotate and mesh.

The meshing of the threads on the rotors, and the close fit of the surrounding housing create one or more sets of moving seals in series between pump inlet and outlet. These sets of seals act as a labyrinth and provide the screw pump with its positive pressure capability.

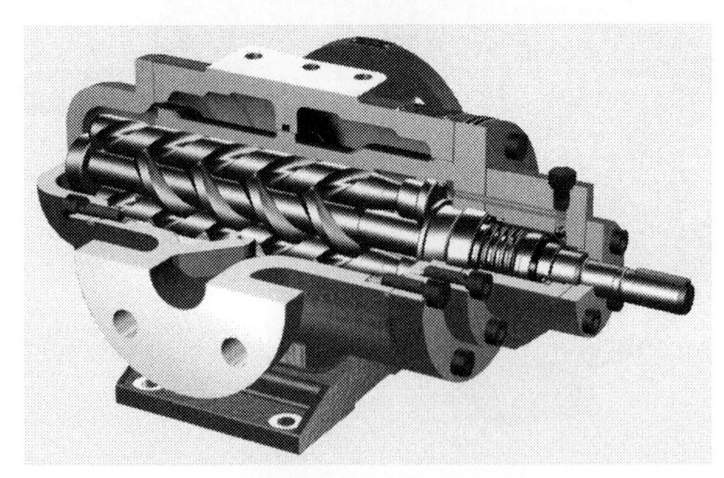

Figure 9.19: Typical screw pump (Reproduced with permission of Roper Pump Company)

The successive sets of seals form fully enclosed cavities which move continuously from inlet to outlet. These cavities trap liquid at the inlet and carry it along to the outlet in a smooth flow.

9.4.5 Progressive cavity pump

This pump has been referred to as a single-end, single-rotor type of screw pump where the pumping elements comprise a single rotor and a stator. The stator usually has a double helical internal thread with a pitch twice that of the single helical stator. This results in two leads on the stator, and one on the rotor.

As the rotor rotates inside the stator, two cavities form at the suction end of the stator, with one cavity closing as the other opens. The cavities progress in a spiral from one end of the stator to the other. The result is a flow with relatively little pulsation, and the shear rates will also be low in comparison to radial pump styles.

The compressive fit between the rotor and stator creates seal lines where the two components contact. The seal lines keep the cavities separated as they progress through the pump with each rotation of the rotor. The elastomeric stator and stainless steel rotor allow the pump to handle large solid particles in suspension and a certain percentage of abrasives.

The manner in which the rotor turns within the stator complicates the mechanical design of PC pumps. As the rotor turns in the stator, the centerline of the rotor orbits about the centerline of the stator. This eccentric motion means the pump must be fitted with universal joints to transmit power from the concentric rotation of the drive shaft to the eccentrically rotating rotor. These joints must transmit torsional and thrust loads. Designs of this drive mechanism range from simple ball-and-pin mechanisms to heavy-duty sealed gear couplings.

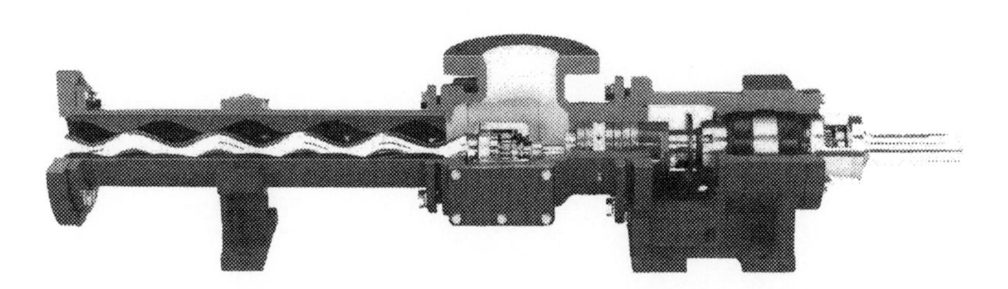

Figure 9.20: Progressive cavity pump (Reproduced with permission of Moyno Inc.)

Pump installation & piping

10.1 Installation

Improper installation of pumps can lead to premature failure and increased maintenance costs. However, when correctly installed and given reasonable care and maintenance, they will operate satisfactorily for a long period of time. The majority of the guidelines detailed in this Chapter apply to a typical horizontal, end suction, single stage centrifugal pump.

10.1.1 Location

When planning the location of the pump, give some consideration to the people who will eventually have to look after that piece of equipment. Wherever possible, all pumps should be located in an area where there is ample space around the pump to provide easy access and working room for routine maintenance. Adequate overhead space should also be available for lifting devices and working clearance. While suction and discharge piping should be as short as possible to minimize friction losses, they should have sufficient space to ensure that correct piping practices can be followed. Further details on pump piping are provided in Chapter 10.2.

10.1.2 Receiving

Upon receipt of a pump on site, the unit should be carefully inspected for damage and checked against the bill of lading. Report any damage or shortages to the carrier's local representative and send a copy to the pump supplier. When uncrating, be careful not to discard any small accessories that may be attached.

10.1.3 Handling

When lifting the pump unit it is essential to use proper lifting techniques and to follow the instruction provided by the pump supplier. It is important to realize that some manufacturers include lifting eyes or eye bolts on various parts of the pump, but that these are designed for lifting only these items and not for lifting the complete pump assembly. A pump should be lifted by positioning the slings around the pump casing and below the bearing housing. When lifting a baseplate mounted assembly, the sling should be under the pump casing and the driver.

10.1.4 Foundation leveling

A proper foundation and grouting can mean the difference between a unit that gives many years of trouble-free service and one that requires constant realignment. It should therefore be everyone's concern that only the best of materials, together with proper design, be used when performing this important function.

The concrete foundation must be sufficiently substantial to absorb any vibration and to form a permanent rigid support for the baseplate. A rule of thumb that is frequently used is that the foundation should be about 5 times the weight of the pump/motor assembly, and approximately 6 inches longer and wider than the baseplate. The anchor bolts are usually of the sleeve type or the 'J' type shown in Figure 10.1 and must be located in accordance with the certified outline drawing of the baseplate.

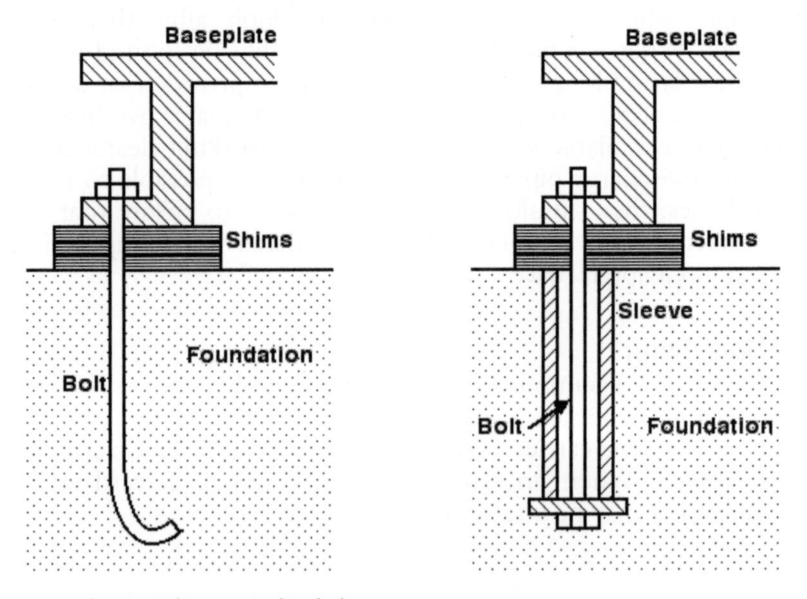

Figure 10.1: Sleeve and J type anchor bolts

Prior to the mounting of the baseplate, the entire surface of the foundation must be leveled by chipping away any defective concrete, leaving it rough, but level. The surface must then be freed of all oil, grease and loose particles. Any waste material stuffed around the foundation bolts must be removed.

While individual practices vary throughout industry, there is a tendency in most leading companies to remove the pump and driver from the baseplate prior to installation. This facilitates leveling of the base and in removing any distortion.

From the underside of the baseplate, clean away any oil, grease, dirt and any coatings that may interfere with complete bonding, or may react with the cement. In addition it is also necessary to avoid trapping air beneath the baseplate when the grout is introduced. This may be achieved by providing one generously sized grouting opening in each bulkhead section of the baseplate, as well as small vent holes in each corner of all such compartments.

10.1.5 Mounting the baseplate

The baseplate should be supported on leveling screws, shims or on metal wedges with a small taper located close to the foundation bolts. All leveling screws and other areas requiring protection from grout spatter should then be covered with a wax to prevent the grout from adhering to them.

The machined mounting surfaces on the baseplate should be checked to within 0.002 inches per foot by adjusting the leveling devices and using a precision level. Ensure the machined mounting surfaces of the baseplate are horizontal, flat and parallel. To prevent stress and distortion of the equipment, all surfaces in the same plane must be within 0.002 inches overall.

When the baseplate is level, check that all support wedges or shims are in full contact with the foundation and baseplate. Tighten the foundation anchor bolts evenly and double-check the level.

10.1.6 Cement based grouting

Build wooden forms around the foundation and saturate the top surface for the necessary period of time as identified by the grout manufacturer. Just prior to grouting, remove excess surface water and any water that may have flowed into the anchor bolt holes.

The temperature of the baseplate, grout and foundation should be kept between 40 and 90°F. during grouting and for a period of at least 24 hours afterwards. Follow the grout supplier's instructions in detail and ensure that the placement of the grout is done quickly and continuously to prevent cold joints and voids under the baseplate. The grout should then be allowed to harden in accordance with the manufacturer's instructions.

It should be noted that vibrations from machines operating close at hand is often transmitted into the foundation of the pump being grouted. These machines should be shut down until the grout takes its initial set, otherwise the bond of the grout may be detrimentally affected. To identify the presence of local vibrations, place a shallow pan of water on the pump baseplate and observe the surface for movement.

10.1.7 Epoxy based grouting

The use of epoxy grout is steadily increasing in a number of industries. The increased bond strength helps to maintain the alignment between the pump and driver, thereby reducing misalignment failures and maintenance costs. Also, as the surface is nonporous, it facilitates cleanup in the event of an accidental spill.

Prior to using an epoxy grout however, the foundation must be clean and dry. The grouting materials should also be stored at temperatures between 70 and 90°F for at least 24 hours before mixing, and it should be poured at temperatures between 40 and 90°F.

10.1.8 Pregrouted baseplates

In recent years some considerable success has been achieved with using pregrouted baseplates to eliminate many of the traditional problems identified with onsite baseplate leveling and grouting. Baseplates are grouted in an upside-down position in the factory which reduces epoxy grout volumetric shrinkage that can otherwise cause baseplate distortion. They have been shown to travel better and arrive at the site flat and aligned, just as they left the factory. They have also been installed on site with the pumps and motors still in place.

10.2 Piping considerations

Piping layout design is an area where the basic principles are frequently ignored, resulting in problems such as hydraulic instabilities in the impeller, which translate into additional shaft loading, higher vibration levels and premature failure of the seal or bearings.

As there are many other reasons why pumps could vibrate, and why seals and bearings fail, the trouble is rarely traced to incorrect piping. The main difficulty is that inadequate piping locates the root cause of many pump failures outside the physical confines of the pump itself, thus making it difficult to detect for the unwary and inexperienced.

It has been argued that piping procedures are not important because many pumps are piped incorrectly, but are operating quite satisfactorily. That doesn't make a questionable piping practice correct, it merely makes it lucky.

Some installations appear as if the pumps have been squeezed into a corner out of the way, and the pipes threaded in and out, without any consideration for fluid flow patterns. This should be strenuously avoided.

When a pump is running, the liquid must arrive at the impeller eye with the right pressure and the smooth uniform flow that is necessary for reliable operation. This depends a great deal on the suction piping design.

10.2.1 Location

The location of the pump relative to its suction source is critical to its ultimate reliability. Every pump should be located as close to its suction source as possible in order to reduce the effect of friction losses on the NPSH available. However it must also be far enough away from the suction source to ensure that correct piping practices can be followed.

These piping practices involve a number of simple rules which, if followed, will eliminate a significant percentage of all potential pump problems.

10.2.1.1 Pipe size

The pipe diameter on both the inlet and the outlet sides of the pump should be at least one size larger than the nozzle itself.

On the suction side it is necessary to reduce the size of the pipe from the suction line to the inlet nozzle. If the inlet nozzle is on a horizontal plane, an eccentric reducer should be positioned with the flat side on top as shown in Figure 10.2. This arrangement eliminates the potential problem of eddy currents in a high point in the suction line that might travel into the impeller eye to the detriment of pump performance.

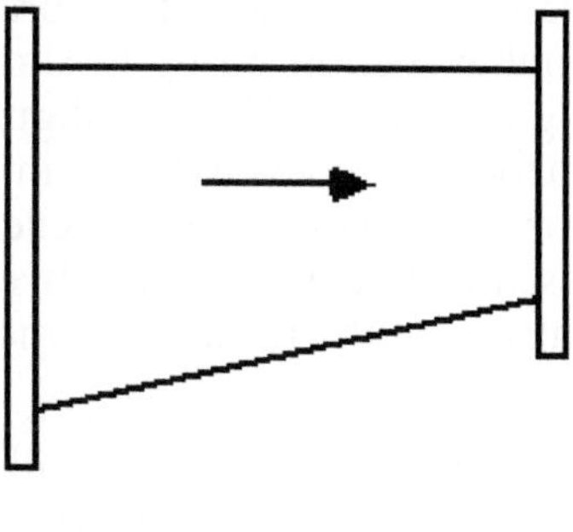

Figure 10.2: Eccentric reducer on suction

A concentric increaser can be used on a vertical discharge and should be bolted to the discharge flange upstream of any valves. This is designed to slow down the velocity of the liquid leaving the pump to an acceptable rate within the system itself and, in particular, through the check valve and isolating valve. The slower velocity (usually lower than 10 ft./sec.) reduces friction losses in the line and minimizes power draw at the pump.

10.2.1.2 Suction elbows

Eliminate elbows mounted on the inlet nozzle of the pump.

Much discussion has taken place over the acceptable configuration of an elbow on the suction flange of a pump. Let's simplify it. There isn't one!

There is always an uneven flow in an elbow, and when one is installed on the suction of any pump, it introduces that uneven flow into the eye of the impeller. This can create turbulence and air entrainment, which can result in impeller damage and vibration.

The only thing worse than one elbow on the suction of a pump is two elbows on the suction of a pump, particularly if they are positioned close together, in planes at right angles to each other. This creates a spinning effect in the liquid that is carried into the impeller and causes turbulence, inefficiency and vibration.

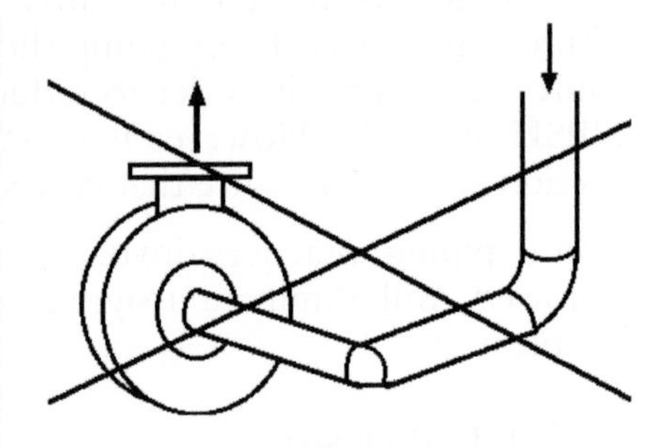

Figure 10.3: Two elbows on pump suction

The problem is compounded to an even greater extent when the elbow is installed in a horizontal plane on the inlet of a horizontal double suction pump as shown in Figure 10.4. This configuration introduces uneven flows into the opposing eyes of the impeller, and essentially destroys the hydraulic balance of the rotating element.

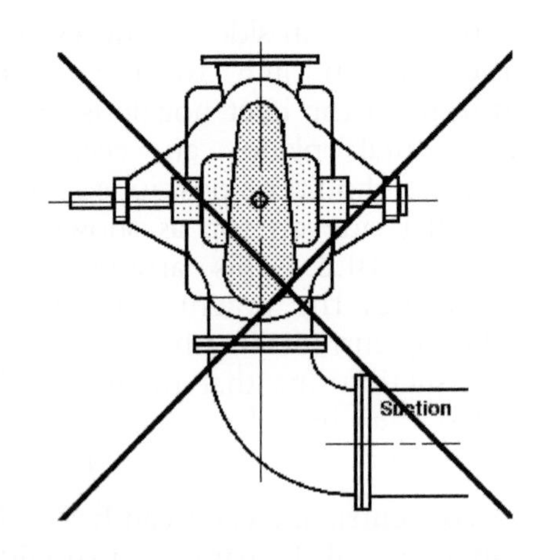

Figure 10.4: D.S. pump with suction elbow

Under these conditions, the overloaded bearing will fail prematurely and regularly if the pump is packed. If the pump is fitted with mechanical seals, the seal will usually fail instead of the bearing, but just as regularly and often more frequently. When it is

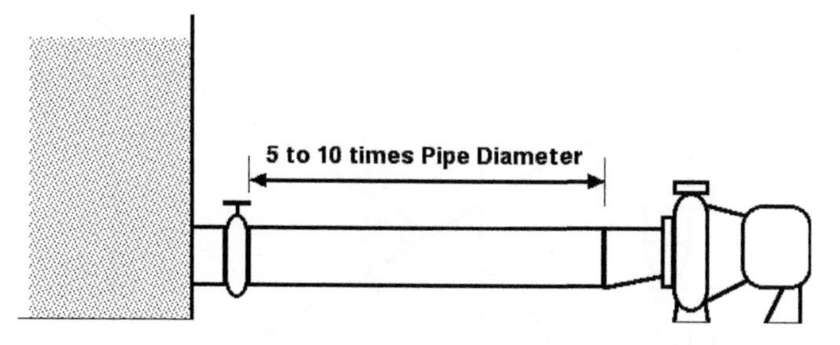

Figure 10.5: Diagram of suction piping

absolutely essential to position an elbow on the inlet of a double suction pump, it must be located on a plane at right angles to the shaft.

10.2.1.3 Straight pipe

Provide the suction side with a straight run of pipe, in a length equivalent to 5 to 10 times the diameter of that pipe, between the suction reducer and the first obstruction in the line.

This arrangement will ensure the delivery of a uniform flow of liquid to the eye of the impeller, which is essential for optimum suction conditions. It allows for any turbulence in the suction tank or the upstream piping to smooth out before impacting the impeller eye. The smaller multiplier will be used on large diameter piping, while the larger multiplier will be needed for the smaller pipe diameters.

10.2.1.4 Air pockets

Eliminate all piping configurations that might introduce an air pocket into the suction side of a pump.

These air pockets can create turbulence and air entrainment, which can result in impeller damage and vibration.

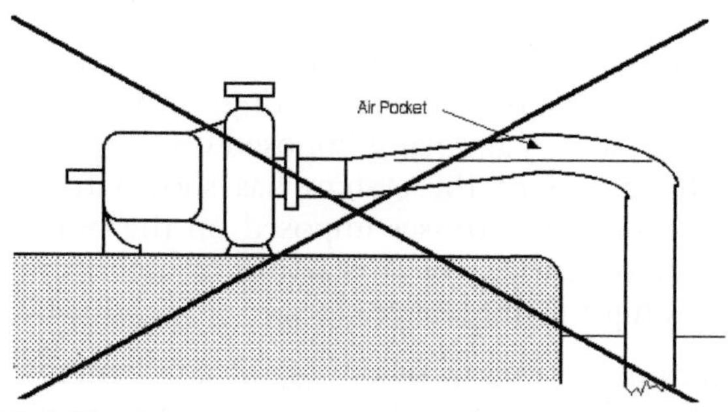

Figure 10.6: Air pocket diagram

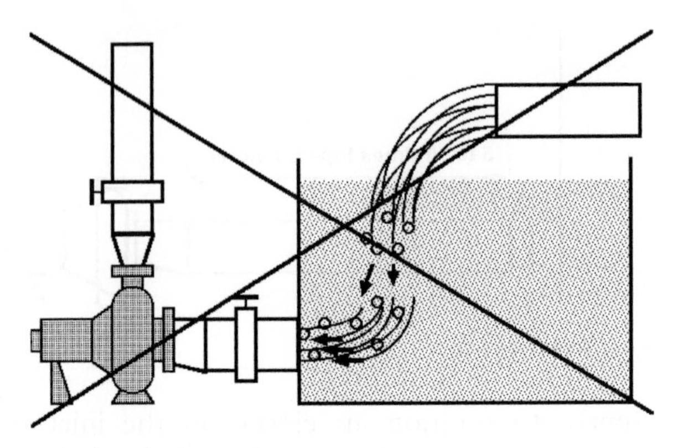

Figure 10.7: Suction tank with air entrainment

10.2.1.5 Suction source design

Eliminate the potential for vortices or air entrainment in the suction source.

If a pump is taking its suction from a sump or tank, the formation of vortices can draw air and vapor into the suction line. This usually can be prevented by providing sufficient submergence of liquid over the suction opening. A bell-mouth design on the opening will reduce the amount of submergence required. This submergence is completely independent of the NPSH required by the pump.

Great care should be taken in the design of a sump to ensure that any liquid emptying into the sump does so in such a manner that air or vapor entrained in the inflow does not pass into the suction opening. Any problem of this nature may require a change in the relative positions of the inflow and outlet if the sump is large enough, or the use of baffles.

10.2.1.6 Pipe strain

Secure the piping in such a way that there is no strain imposed on the pump casing.

Piping flanges must be accurately aligned before the bolts are tightened and all piping, valves and associated fittings should be independently supported without any strain being imposed on the pump. Building the piping from the pump to the system has shown to be beneficial in reducing pipe strain. Any stresses imposed on the pump casing by the piping reduce the probability of satisfactory reliability and performance, and can result in high maintenance costs.

Laser pipe alignment tools provide significant benefits to the pump user. Mounted on the pump or the piping, such alignment tools are an effective way to prevent pipe strain during initial installation, which is exactly the right time to stop it.

The effects of pipe strain in a conventional process pump can also be transmitted to the shaft coupling. With the accuracy of laser alignment technology, pipe strain can be measured much in the same way as soft foot and should be integrated into the shaft alignment program. By loosening the pipe connection after a correct shaft alignment has been performed, any movement at the coupling will indicate the existence of pipe strain and can therefore be corrected.

As there is an exception to almost every rule, the A.P.I. 610 Pump Specification used in the petroleum industry identifies a maximum level of forces and moments that may be imposed on the pump flanges. These must be acceptable to any pump being sold into that, or a related, industry using that specification. As a consequence, all API pumps are of a much more robust and heavier design than their ANSI size equivalents.

In high temperature applications, some piping misalignment is inevitable owing to thermal growth during the operating cycle. Under these conditions thermal expansion joints are often introduced to avoid transmitting any piping strains to the pump. However, if the end of the expansion joint closest to the pump is not anchored securely, the pipe strain is passed through to the pump, thus defeating the objective of the expansion joint.

To ensure even gasket compression when tightening up the flange nuts and bolts, always tighten the bolts diametrally opposed to each other in an alternating pattern.

10.2.1.7 Pipe fittings

The Suction Valve has only one function, and that is to isolate the pump from the system during maintenance. It should never be used to throttle the flow rate as this will cause cavitation and suction recirculation in the eye of the impeller.

The Suction Strainer may be required during the commissioning stage, or other emergency conditions. In such cases, a strainer with a net area of at least 5 to 6 times the size of the suction line is usually recommended. However, in view of their function, they will gradually build up an increased resistance to an even flow to the pump. Consequently, they should be avoided whenever possible during normal system operation. Any straining of particles from the pumpage should be carried out downstream of the pumps, and the pumps selected to be able to pass such particles.

The Discharge Valves usually constitute a non-return check valve and

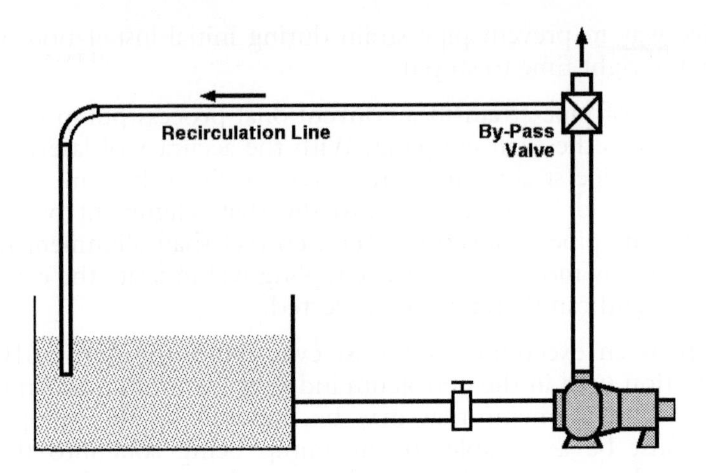

Figure 10.8: Recirculation pipe diagram

some kind of isolating valve. As the check valve protects the pump from being driven backwards by returning fluid when the pump is shut down, it should be located closest to the pump discharge valve. In addition to isolating the pump for maintenance, the isolating valve may be opened slowly when the pump is started up on an open system, and thus prevent the pump from running too far out on the pump performance curve while the system is being filled. Both discharge valves should be located in the system downstream from the concentric pipe increaser to prevent unnecessary turbulence at the valve.

A By-Pass Valve may be necessary to regulate the discharge flow if prolonged periods of low flow operation are expected due to variable system demand. It may also prevent an excessive number of pump stops and starts which is particularly important with larger motor drivers.

As shown in Figure 10.8, the recirculated flow should direct excess flow back to the suction source of the pump. It must not be directed back to the suction of the pump where excessive line turbulence may result.

10.3 Alignment

Industry has progressed well beyond any discussion of the statement that good alignment is essential to safe and trouble-free operation of rotating equipment. The only questions that still remain in some areas are; 'what is alignment?', and, 'what is good alignment?'.

When we discuss 'Alignment' in the pump industry, we are discussing either 'Piping Alignment' (which we reviewed in Chapter 10.2.1.6) or 'Shaft Alignment'. It must be noted that the term 'Coupling

Alignment' is a misnomer. We are not concerned about bringing the coupling halves into alignment, we are only interested in ensuring the shafts of the pump and its driver will rotate on a common axis. If the shafts are not coaxial, the resulting moments will increase the forces on the pump shaft and bearings, causing accelerated wear and premature failure.

10.3.1 Shaft couplings

That does not mean that we should ignore the coupling. There are only two types of couplings, both of which are capable of operating with some degree of misalignment. One type accommodates misalignment through the sliding of one element over another, while the other accommodates misalignment through the flexing of one or more components.

Any couplings that operate through the sliding of one element over another will not only require lubrication, but may prefer some minimal movement between the elements in order to keep the surfaces lubricated. In other words, these couplings might even last longer with some minimal shaft misalignment because, without any relative motion, the lubricant will be expelled from the spaces between the element surfaces, thus causing premature wear.

Those couplings that accommodate misalignment through flexing as shown in Figure 10.10, will last longer and work better at zero misalignment as there are no forces being transferred through to the pump shaft and bearings.

What should be recognized however is that all couplings resist being operated when misaligned, and the restoring forces and moments

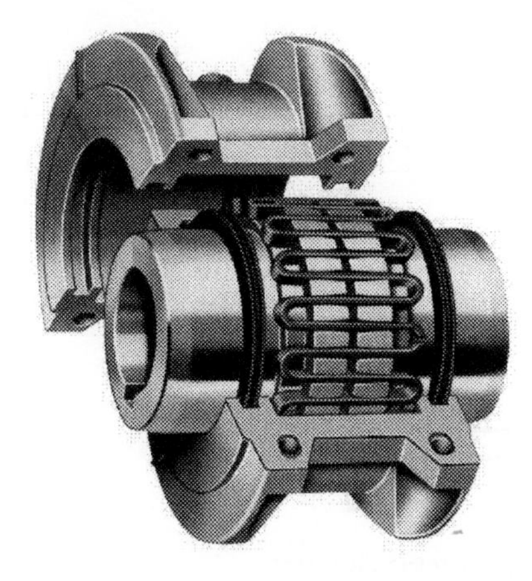

Figure 10.9: Gear type coupling (Reproduced with permission of the Falk Corporation)

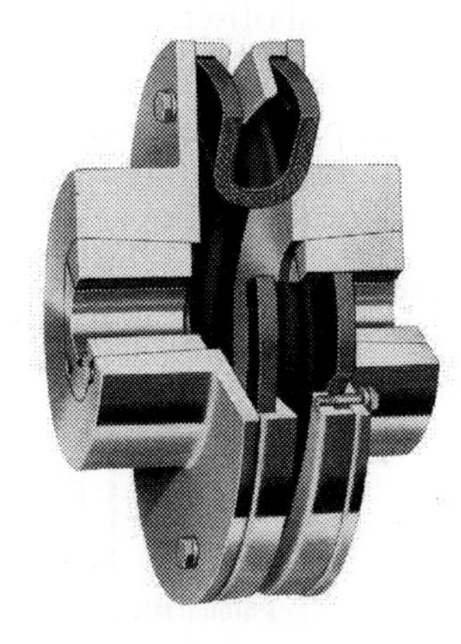

Figure 10.10: Rubber type coupling (Reproduced with permission of the Falk Corporation)

involved in this resistance can damage bearings, seals and even the shaft itself. It should also be noted that the restoring forces are proportional to misalignment. In other words the larger the misalignment of the shafts the greater the forces on the bearings, etc. Consequently, the shafts of the pump and driver should be aligned as closely as possible to provide optimum reliability of the pump.

In most pump installations, it is accepted that perfect shaft alignment is unlikely throughout the operating cycle. In such conditions, the coupling selection should be able to accommodate the maximum amount of the misalignment anticipated. This should be confirmed with the supplier as even flexible couplings have limitations which are often ignored and result in premature bearing failure and unreliable operation.

10.3.2 Shaft offset and angularity

Alignment occurs when two lines that are superimposed on each other, form a single line. Misalignment is a measure of how far apart the two lines are from forming that single line. The two lines we are concerned with here are the centerlines of the pump shaft and the driver shaft. In one condition, the two lines can be parallel with each other, but at a constant distance apart. This is referred to as Offset or Parallel Misalignment. In the other, one line will be at an angle to the other, and is referred to as Angular Misalignment.

Parallel misalignment can be considered as the distance between the driver shaft centerline and the pump shaft centerline at any given point along the length, and this misalignment can happen in any plane. Consequently, it is worthwhile to take the necessary measurements on the top and on the bottom for vertical offset and also on each side for the horizontal offset.

Angular misalignment refers to the difference in slope of the two shafts. If the pump, base and foundation have been properly installed, the shaft centerline of the pump can be considered as level and therefore, as the reference or datum line.

The slope of the driver shaft can be calculated by determining the offset measurement at two different points, subtracting one from the other,

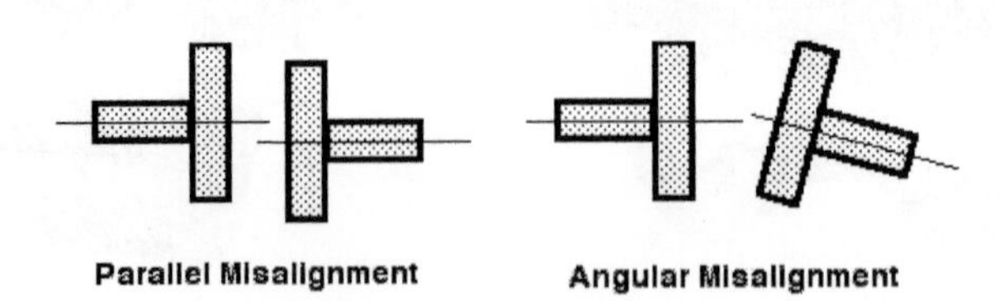

Figure 10.11: Shaft offset and angular misalignment

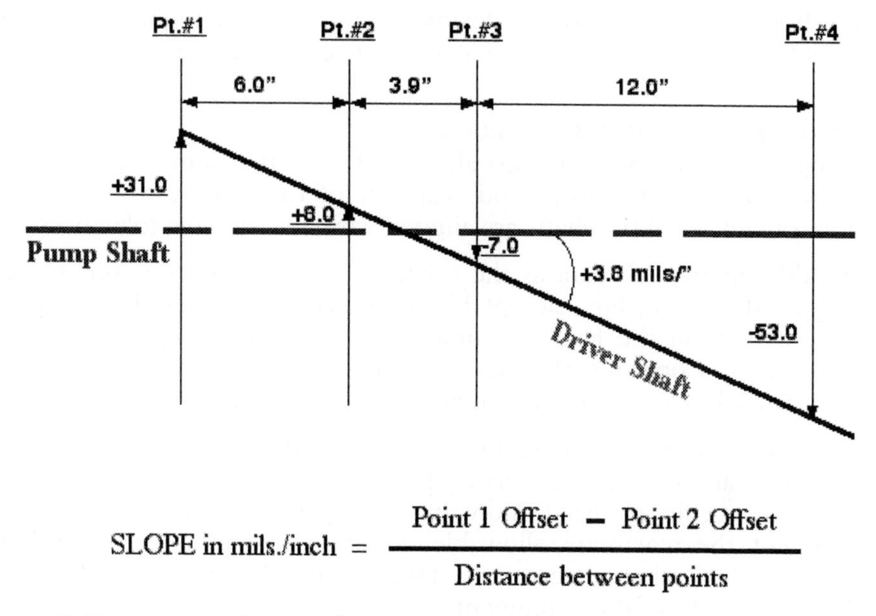

$$\text{SLOPE in mils./inch} = \frac{\text{Point 1 Offset} - \text{Point 2 Offset}}{\text{Distance between points}}$$

Figure 10.12: Angular misalignment diagram

and then dividing the result by the axial distance between the two points. This misalignment should be measured and calculated in both the vertical and horizontal planes.

10.3.3 High temperature corrections

When a foot mounted process pump is required to operate at elevated temperatures, some adjustment will be necessary to allow for the thermal growth that takes place between the cold condition and the high operating temperatures. As the pump heats up, the shaft centerline will be moved up by the thermal growth of the pump, creating an offset with the motor shaft.

One method of handling this situation is to misalign the motor by the amount of growth anticipated from the pump prior to starting the pump. Most pump manufacturers can provide the cold setting figures corresponding to the higher operating temperatures. This will require the pump and motor shafts to run in a misaligned setting until the pump is fully up to temperature, by which time, the expansion of the pump will raise the pump into position to align with the motor.

A second method is to start the pump and motor following a cold alignment without any adjustment. As the pump heats up and expands, it will gradually move up, out of alignment with the motor. When the pump is fully up to temperature, the unit is stopped and hot alignment takes place.

For both of these methods, a flexible coupling will be required that will be able to accommodate the total amount of misalignment anticipated.

10.3.4 Typical acceptance values

Bringing the motor shaft into alignment with the pump shaft usually involves moving the front and rear feet of the motor, vertically and horizontally, until the shafts are aligned within acceptable tolerances.

In addition to such data as the speed of rotation, horsepower, spacer length, shaft size, etc., acceptable alignment tolerances depend to a large extent on the level of pump reliability that is expected by the pump user. Consequently, every end user should develop their own acceptance levels that provide their desired outcomes.

The tolerances shown in Figure 10.13 are offered as guidelines only, but they can be used as a starting point for developing tolerances that will be specific to each individual company or equipment. They represent the maximum allowable deviation from the desired value, whether that value is zero or a targeted misalignment to allow for thermal growth of the equipment.

10.3.5 Run out

With the coupling disconnected, mount the magnetic base of the dial indicator to the motor half coupling, position the indicator on the pump half coupling, and center the indicator plunger. Rotate the pump shaft until the dial indicator reaches a maximum travel and zero the dial indicator. Rotate the pump shaft again until the dial indicator reaches a maximum value. This shows the amount of run out.

Condition	R.P.M.	Tolerances
Parallel Misalignment	3600	0.002 ins
	1800	0.004 ins
	1200	0.005 ins
Angular Misalignment	3600	0.004 ins/inch
	1800	0.006 ins/inch
	1200	0.008 ins/inch

Figure 10.13: Alignment tolerances table

If the run out on the pump side is in excess of the normally acceptable limit of 0.002 inches, the pump shaft run out should be checked as above, except with the dial indicator applied to the shaft. If the shaft run out is 0.001 inch or less, then the shaft can be considered acceptable, but the coupling is eccentric. If however, the shaft run out is greater than 0.001 inch, the shaft should be straightened or replaced. By switching the position of the dial indicator, the driver shaft can be checked in the same manner with the same limitations.

10.3.6 Soft foot

To check for soft foot prior to alignment when there are no shims under the motor feet, start out by trying to fit a 0.005 inch shim under each foot. If the shim will fit under a foot, make up the gap by gradually increasing the shim thickness until a tight fit is achieved. If shims do exist already, ensure that there are no more than 4 shims in any one location. If there are, consolidate them by using thicker shims. Check at each foot for loose shims and make up the gap by gradually increasing the shim thickness until a tight fit is achieved at all feet.

A final soft foot check should be performed only after any vertical angular misalignment has been corrected. When that has been achieved, mount the dial indicator to contact the foot to be checked and set the indicator to zero. Loosen the hold down bolt on that foot and record the dial indicator reading, and then retighten the hold down bolt. Repeat this process with all four feet.

Soft foot conditions in excess of 0.002 inch should be corrected by adding shims to the foot with the largest soft foot value. Note that excess shims will result in increased soft foot at the other feet. Check other feet and make any necessary corrections.

10.3.7 Alignment methodology

While dial indicators are still a viable method of establishing shaft alignment, laser alignment systems are now providing increased accuracy that reduces maintenance costs while improving pump reliability. In the current industrial workplace where fewer people are expected to do more, laser alignment systems not only reduce the time required to achieve a high level of alignment accuracy, but they do so without the need for any expertise in mathematical graphing and calculation.

10.3.8 C–Flange adapter

One of the major benefits of having a C–Flange mounting arrangement where the pump and motor are fitted to a common adapter. With this design shaft alignment is not required, as the rabbeted fits of the motor to the adapter and the adapter to the pump bearing housing automatically aligns the shaft to within acceptable limits of the pump manufacturer. A number of pump manufacturers have now made this arrangement available in limited sizes.

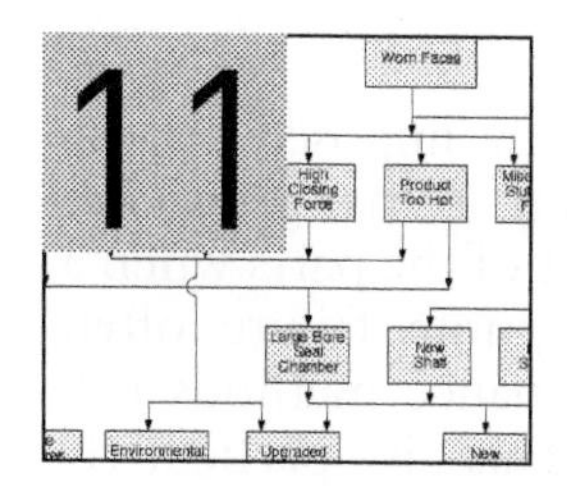

11 Troubleshooting

11.1 Skill and experience

One of the major problems facing industry today is the limited number of people who have sufficient skill and experience to diagnose and rectify the basic problems plaguing centrifugal pumps. The other major difficulty is that the same lack of skill and experience is creating many of these problems in the first place.

A detailed evaluation of a pump problem requires a depth of knowledge which usually surpasses that to which most people are ever exposed. For example, most pump engineers, operators and maintenance people develop their knowledge base from the same 'school of hard knocks'. While this on-the-job type of training has much to commend it, it unfortunately exposes the pupil to the opportunity of learning other people's mistakes and misconceptions. At best, it only teaches what is necessary to execute a particular job function in exactly the same manner as it was previously performed – good or bad!

A system engineer may have learned how to size a pump based on the operating parameters of the system. However, if that pump is unable to withstand the effects of certain installation or operating anomalies which may occur in that plant, the reliable life of the pump could be detrimentally affected.

A purchasing manager may be faced with the responsibility of buying equipment within a capital cost budget which has no mechanism for the evaluation of long term operation and maintenance costs. Consequently the 'most economical' pump purchased may result in frequent and repetitive failures which could quickly exceed the initial cost difference and even the total price of the pump itself.

Operations personnel are often required to 'tune' the system to provide the desired output of product. In this endeavour, they may be faced with the necessity of throttling back the pump discharge valve, and this can result in a variety of hydraulic and mechanical problems in the pump.

The ramifications of all such situations are generally imposed on the maintenance department. Unfortunately, the training in that group has traditionally been limited to the physical change-out of the parts when a breakdown occurs. As the underlying cause of pump failure often extends well beyond the failed item, these maintenance methods will effectively reinstall the same old problem. This is particularly concerning when we realize that over 80% of all pump failures tend to manifest themselves at the mechanical seal or the bearings, which then act in a manner similar to a fuse in an electrical system.

When a fuse in an electrical system fails, it doesn't mean there is anything wrong with the fuse! In fact we understand that the problem is almost always somewhere else in the system. In spite of this, when a seal or bearing fails, we rarely look for the real problem. Instead, we simply replace the offending part. While that will occasionally solve the problem, the simple change-out of a seal or bearing rarely provides long-lasting relief from the problem. Consequently, we have to review the two different types of pump problems – those that are either operational or reliability in nature – as well as some sources of these problems, and some tools for identifying their underlying causes.

11.2 Operational problems

This is the type of problem where the pump simply doesn't produce the hydraulic results for which it was intended. A typical example of this occurs when the centrifugal pump isn't pumping enough liquid through the system. While this is often blamed on an inadequate pump, there are many other conditions that could be to blame, such as is shown in Figure 11.1. The problem may be a blocked inlet line or air entering the inlet line or even, in an appropriate case, a lack of priming.

11.2.1 Cavitation problem solving

Another example is when the pump is operated so far away from its design point that it begins to vibrate as a result of a variety of hydraulic conditions identified on Figure 11.2.

However, the biggest problem in all this is to identify which of three hydraulic conditions we are faced with when the common symptoms of noise and vibration are experienced.

This is accomplished by the throttling of the discharge valve which reduces the flow through the pump and creates three possible scenarios.

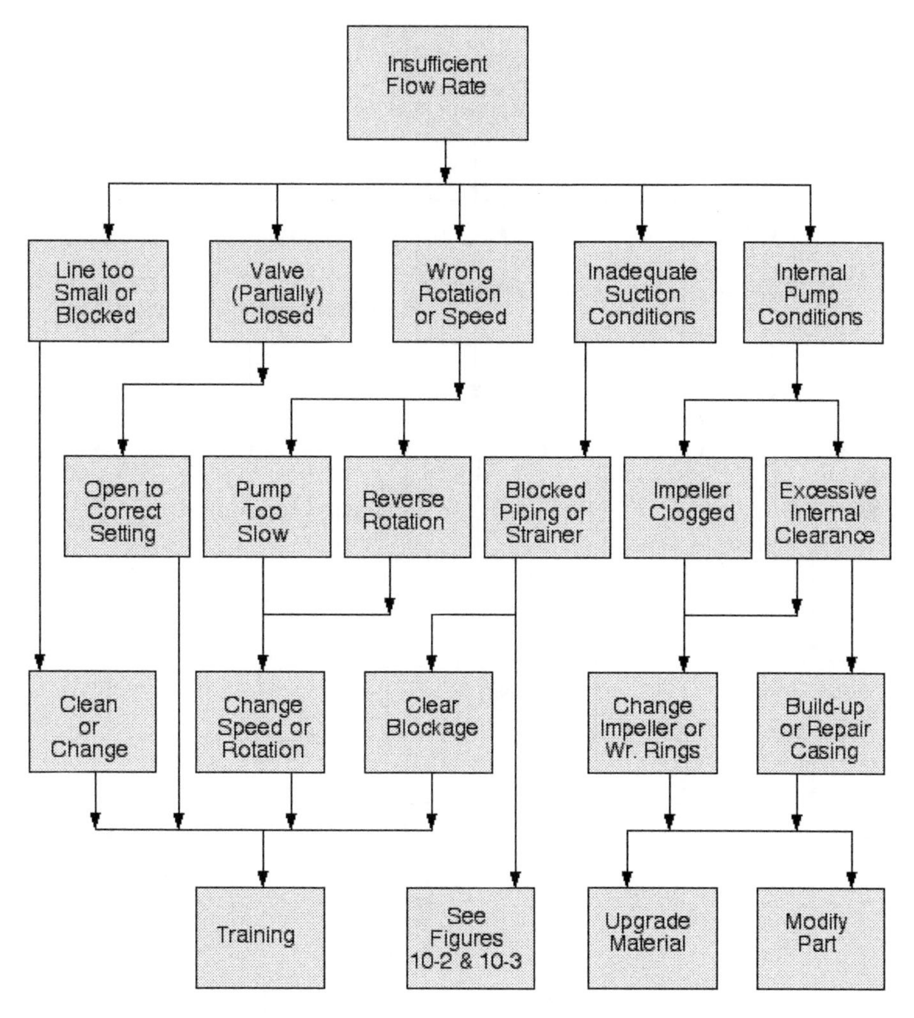

Figure 11.1: Insufficient flow rate chart

- The noise and vibration will grow quieter and perhaps even die away completely
- The noise and vibration will get worse
- Little or no difference is experienced

With the first result, the pump will be operating at a lower flow where a lower level of NPSH is required, and the quieter, smoother operation identifies that Cavitation is being eliminated.

If the noise and vibration gets worse, it indicates that the pump is moving into a worsening condition of low flow which demonstrates a problem with Recirculation.

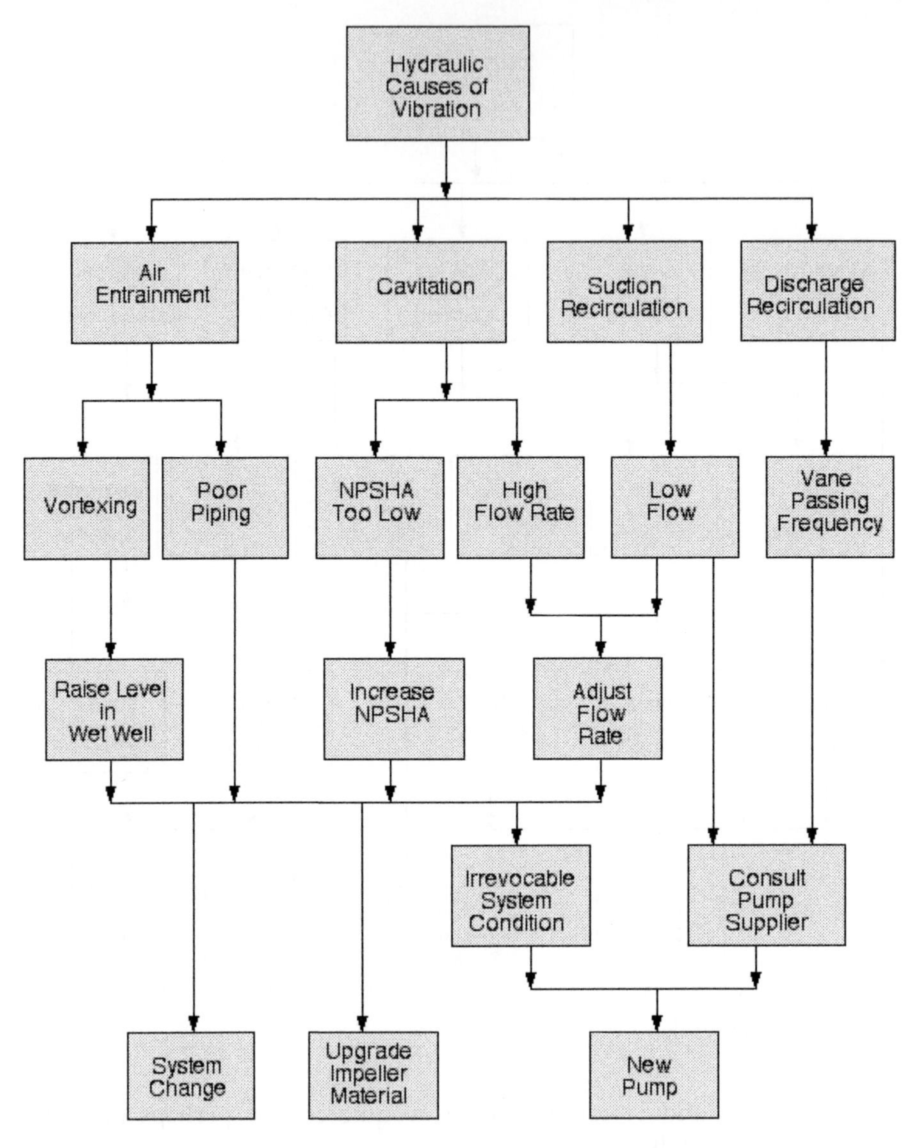

Figure 11.2: Hydraulic causes of vibration chart

When little difference is experienced, that indicates an Air Entrainment problem that is not immediately susceptible to changes in flow rate.

11.2.2 Uncurable conditions

While the various possible cures for these hydraulic problems are clearly identified, it is recognized that, on many occasions, the appropriate cure cannot be implemented. For example, it is difficult to raise the level of the liquid in the wet well, if that wet well happens to be an ocean, lake or river. Consequently it may be necessary to live with the problem.

That does not mean to say that it is necessary to live with the ramifications of the problem. If we can't cure the problem, we can frequently make a difference to the damaging effects of the symptom.

For example, with a pump that is cavitating, there are four identifiable symptoms;

- A peculiar rumble/rattling noise,
- high vibration levels,
- a pitted impeller, and
- a slight reduction in total head.

The use of ear plugs is about the only relief that can be experienced from the noise level, and a slight increase in impeller diameter or rotational speed will take care of the total head.

The extent of the pitting damage can be frequently aleviated with the use of a harder material for the impeller. For example, a stainless steel impeller has been known to last 4–6 times as long as a bronze impeller in the same adverse conditions.

The high vibration levels can sometimes be reduced with the use of a stronger shaft as will be discussed in Chapter 11.4.2.2.

While none of these options eliminates the problem, they may allow the plant personnel to live with the reduced detrimental effects of the symptoms.

11.3 Reliability problems

Reliability problems bring into question the length of time the pump can be expected to continue running. While this would include the damaging aspects of a cavitation type of problem as discussed in the previous paragraph, a more typical example is when a pump is vibrating as a result of a variety of mechanical conditions as identified in Figure 11.3.

Although there have been almost 100 different problems identified with centrifugal pumps, many of which have more than one solution – sometimes they have two, three, or even more solutions – it is interesting to note that there are much fewer solutions. In fact, if you scrutinize the accompanying tables, it is evident that there really are only 6 basic solutions to all pump problems.

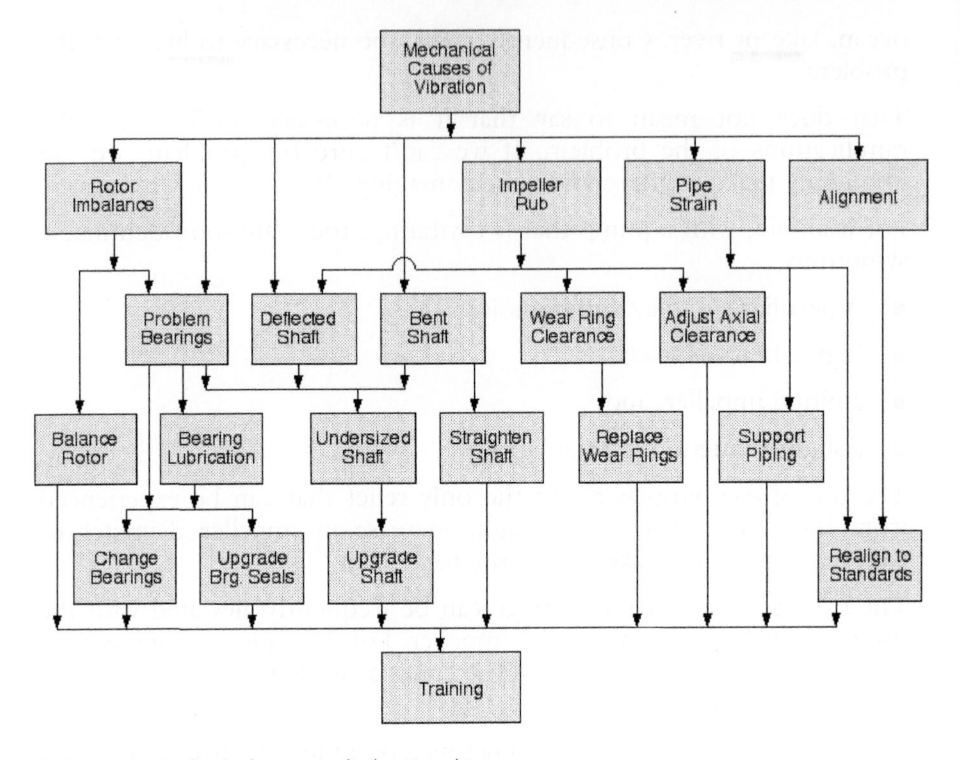

Figure 11.3: Mechanical causes of vibration chart

- Better Sealing Devices
- Component Modification
- Upgraded Materials
- System Change
- New Pump
- Personnel Training

Of these six, the most significant and far-reaching one is that of providing adequate personnel training. This means that everyone who has anything to do with the sizing, selection, installation, operation and maintenance of the pump, is fully skilled and experienced in their own particular function. They also need to be aware of how their actions impact the actions of others, as well as the overall reliability of the pumping system.

11.4 Failure analysis

As there are only a few symptoms with which to recognize a troubled pump, and the key to failure analysis lies in understanding how the combinations of symptoms identify the underlying cause of the problem.

11.4.1 Speed of problem occurrence

An effective troubleshooting tool will always begin with the question, 'How fast did it show up?'. If the problem has only suddenly appeared, it is likely to have a different cause than a similar problem that has been developing over time. It is also fairly obvious that a sudden appearance of the problem is probably caused by a sudden change in the condition that created the problem. Therefore it is highly unlikely that such a problem can be attributed to normal wear and tear. It is much more probable that an inappropriate action has taken place quickly.

The exception to that concept is where wear gradually takes place until the point at which failure suddenly occurs. In such an event however, the wear is usually indicated by a gradual reduction in performance until the breaking point is reached; thus providing some prior notification of imminent failure. This type of condition underscores the need for constant measurement of performance as it relates to temperature, pressure, flow, vibration and power draw.

11.4.1.1 Excessive power consumption

For example, some of the more frequent causes of excessive power consumption are as follows:

- Flow rate is higher than expected
- System pressure is lower than expected
- Pump speed is too high
- Impeller diameter is too big
- Rotating element is binding somewhere
- Impeller is rubbing on the casing
- Wear rings are touching
- Packing is too tight
- High load on bearings
- Pumpage has changed to higher density

Many of these causes can happen abruptly, while others can occur gradually.

11.4.2 Frequency of problem occurrence

A typical example of this problem is when a mechanical seal in a particular pump fails every six months, regardless of the type of seal used in that pump. Maintenance may have tried many different models, types and face material combinations, but the seal fails with the same frequency every time. As it is logical to expect different seals to last different periods of time between failures, it becomes evident fairly quickly that this is a situation where the seal is simply acting as the 'fuse' in the system. Consequently, the underlying problem is obviously elsewhere in either the pump or the system.

This is where skill and experience come in to play. The condition described above is one where the experienced troubleshooter would immediately consider either the pump shaft or the piping arrangement, depending on the pump model in question.

11.4.2.1 Hydraulic imbalance in a double suction pump

For example, a horizontal, double-suction pump may be fitted with a 90° elbow mounted on the suction nozzle in such a way that the line leading to the elbow is parallel with the axis of the pump shaft as shown in Figure 11.4. When the liquid sweeps round the elbow it is centrifuged out towards the long radius and feeds the eye on one side of the impeller, effectively starving the opposing eye. This creates an imbalance of the liquid in the pump casing that can cause an excessive axial thrust to be imposed on the impeller.

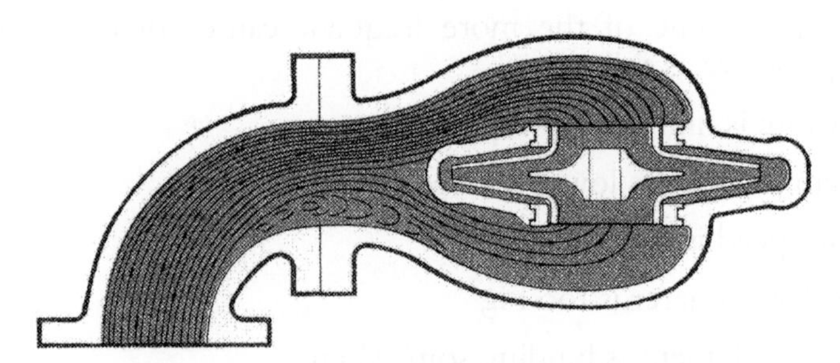

Figure 11.4: D/S pump with suction elbow (Reproduced with permission of Goulds Pumps, ITT Industries)

The normal outcome of such an arrangement is a consistently frequent failure of the mechanical seal or (when packing is fitted) the bearing, at the end of the shaft closest to the suction source. Such failure will normally occur at approximately 6 month intervals regardless of the type of seal or bearing that is installed.

11.4.2.2 Undersized shaft in an end suction pump

In a horizontal, end suction centrifugal pump, frequent and regular seal failure with different seals indicates an undersized shaft that is being subjected to excessive deflection.

Diam. "D" (ins.)	Length "L" (between Radial Bearing & Impeller centerlines)								
	6.0	6.5	7.0	7.5	8.0	8.5	9.0	9.5	10.0
1-1/8	135	172	214	264	320	384	456	536	625
1-1/4	88	112	140	173	210	251	299	351	410
1-3/8	60	77	96	118	143	172	204	240	280
1-1/2	43	54	68	83	101	121	144	169	198
1-5/8	31	39	49	61	73	88	105	123	143
1-3/4	23	29	37	45	55	56	78	91	106
1-7/8	17	25	28	34	41	50	59	69	81
2	14	17	21	26	32	38	46	54	63
2-1/8	11	14	17	21	25	30	36	42	49
2-1/4	8	11	13	16	20	24	28	33	39

Figure 11.5: Shaft slenderness ratio chart

The same thing is true of a packed pump which cannot maintain a minimal amount of leakage for any length of time and seems to be constantly leaking excessively regardless of the amount of time and expertise that is spent on minimizing the leakage. This problem is frequently blamed on the last individual that repacked that pump, or even on the type of packing that is used; resulting in many different packing styles being tried. The underlying source of the difficulty is also an undersized shaft that is being subjected to excessive deflection.

The table shown in Figure 11.5 identifies the values of the Shaft Slenderness Ratio (SSR) for varying sizes of pump shaft diameters and the distances between the impeller and the radial bearing. This is a dimensionless number that is derived from the equation shown in Chapter 2.5.2 for calculating shaft deflection. The purpose of this SSR is to provide a relative value for the condition of the overhung section of the shaft in an end-suction pump.

The weakness will escalate as the shaft diameter decreases and the overhung length increases. Reducing the value will therefore improve the strength of the pump shaft.

As an example, consider a pump with an overhung length of 8 inches and sleeve diameter of 1 5/8 inches. This reveals a value of SSR of 73 which, under most operating conditions, would be considered perfectly adequate. Unfortunately this would only be appropriate if the shaft sleeve were shrunk onto the shaft where it contributes to the overall strength of the shaft.

If, however, the sleeve is a hook type or keyed to the shaft, it actually detracts from the shaft strength. Under such conditions, the SSR is calculated with the use of the shaft diameter underneath the sleeve at, say, 1 3/8 inches in diameter, revealing a value of SSR of 143 which is almost double the previously calculated value. This identifies a much weaker shaft that is almost twice as susceptible to deflection in the event of a hydraulic upset condition.

The good news is that this can be quickly corrected by eliminating a shaft sleeve from the pump and using a shaft with a 1 5/8 inch diameter in that area, thus reestablishing the shaft strength.

It must be noted that high levels of SSR will only be a problem if the pump is required to operate under conditions where high radial loads will be in effect. For further details on these radial loads, please refer to Chapter 2.5.1.

11.5 Failure modes

The following lists the major parts of a pump and the failure conditions that tend to show up in these parts, together with a guide towards the possible underlying causes of the failures.

This list should not be considered as being complete, as every industry, every plant and almost every pump, has its own set of peculiarities that show up in pump failure. Consequently, the items and recommended actions identified should be considered only as a guide towards the achievement of pump reliability.

11.5.1 Impeller

Impeller clogging

- Check pump operation
- Check for change in pumpage
- Check impeller clearance
- Change to non-clog impeller design

Impeller imbalance

- Balance impeller

- New impeller

Corrosion problems

- Upgrade the material selection
- Check change in pumpage

Erosion problems

- Upgrade the material selection
- Check change in pumpage

Cavitation damage

- See Figure 11.2

Suction Recirculation

- See Figure 11.2
- Check the specific speed of impeller

Discharge Recirculation

- See Figure 11.2
- Check for vane passing frequency

Air entrainment

- See Figure 11.2
- Check configuration of inlet piping
- Check arrangement of suction source

Temperature problems

- Check impeller axial clearance
- Check wear ring clearance
- Check for change in pumpage

Impeller seizing

- Check impeller axial clearance
- Check wear ring clearance

11.5.2 Wear rings

Corrosion problems

- Upgrade the material selection
- Check change in pumpage

Erosion problems

- Upgrade the material selection
- Check change in pumpage

Bent shaft

- Install a new shaft

Low flow operation

- See Figure 11.3
- Increase the shaft strength. (See 11.4.2.2)

Bearing failure

- See Chapter 11.8

Excessive wear

- Eliminate pipe strain
- Improve shaft alignment

Galling

- Eliminate pipe strain
- Upgrade the material selection
- Check wear ring clearance
- Skills training needed

11.5.3 Shaft

Wear from Seal/Packing

- Change to non-fretting mechanical seal
- Upgrade the material selection
- Use less abrasive packing
- Control the solids in pumpage

Oil Seal fretting

- Change to bearing isolators

Damaged by seal removal

- Change to non-fretting mechanical seals that don't require sleeves

Corrosion

- Upgrade the material selection

11.5.4 Sleeve

Wear from Seal/Packing

- Change to non-fretting mechanical seal
- Upgrade the material selection
- Use less abrasive packing
- Control the solids in pumpage

Damaged by seal removal

- Change to non-fretting mechanical seals that don't require sleeves

Corrosion

- Upgrade the material selection

11.5.5 Bearings

Inadequate lubricant

- Check lubricant supply
- Change to oil sight glass

Rotor imbalance

- Balance rotor

Lubricant contamination

- Change to bearing isolators
- Remove breather cap and plug connection
- Paint inside of bearing housing

Overheating

- Check fits on shaft
- Check fits on housing
- Check for adequate lubrication
- Use cooling coil

Shaft misalignment

- Skills training needed
- Use a C-flange motor

Shaft deflection

- Check for low flow. (See Figure 11.3)

- Increase the shaft strength. (See 11.4.2.2)

Poor installation

- Skills training needed
- Use a C-flange motor
- Check pump base and foundation
- Eliminate pipe strain
- Correct shaft alignment
- Eliminate belt drives

Premature failure

- Clear plugged oil return holes
- Correct the bearing selection
- Check shaft loading
- Minimize vibration. (See Figures 11.1 and 11.2)

11.5.6 Packing

Worn sleeve

- Change to a non-fretting mechanical seal
- Upgrade the material selection
- Use less abrasive packing
- Control the solids in pumpage

High temperature

- Upgrade material selection
- Increase cooling effect of flush
- Change to a mechanical seal

Chemical attack

- Upgrade the material selection
- Change to a mechanical seal

Over-tightening gland

- Skills training needed

High pressure

- Check pump operation for low flow

- Upgrade the material selection
- Change to a mechanical seal

Shaft deflection

- Check for low flow. (See Figure 11.3)
- Increase the shaft strength. (See 11.4.2.2)

Shaft run-out too high

- Install a new shaft

Worn stuffing box

- Install a new stuffing box

Mis-aligned lantern ring

- Repack the pump properly
- Replace stuffing box if necessary
- Skills training needed

Too many rings of packing

- Repack the pump properly
- Skills training needed.

Wrong size of packing

- Repack the pump properly.
- Skills training needed.

Rings incorrectly cut

- Repack the pump properly
- Skills training needed.

Poor start-up procedures

- Knowledge training needed

Badly installed packing

- Repack the pump properly
- Skills training needed

11.5.7 Mechanical seal – general

Temperature problem

- Improve environmental controls

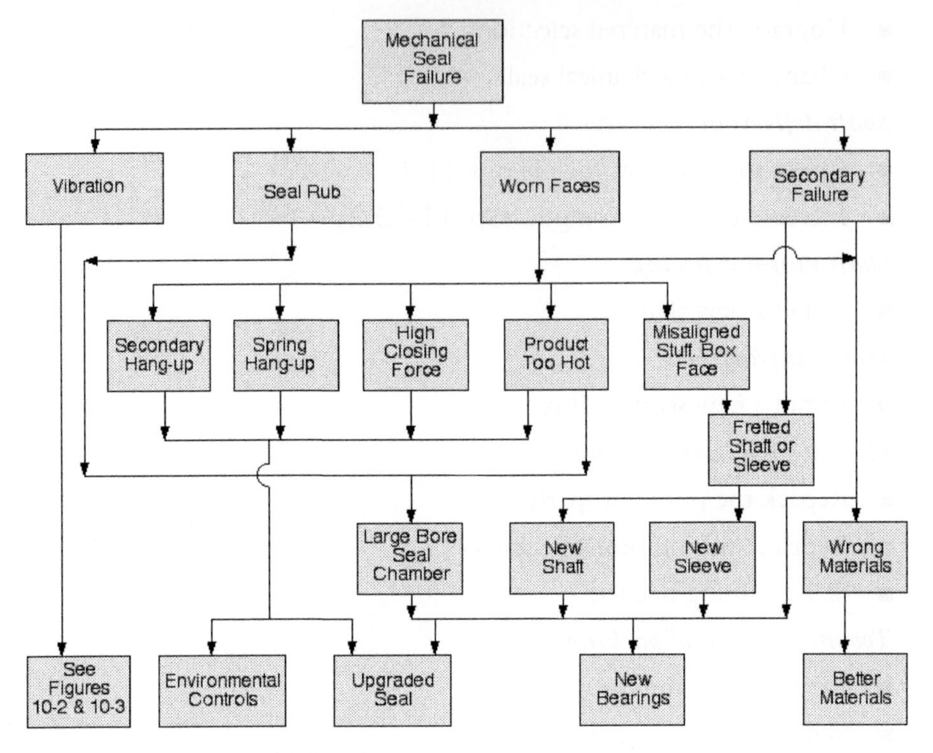

Figure 11.6: Mechanical seal failure chart

- Upgrade the material selection
- Check for low flow. (See Figure 11.1)
- Change to large bore seal chamber
- Check for change in pumpage

Pressure problem

- Improve environmental controls
- Check pump operation for change in flow rate

Corrosion problem

- Upgrade the material selection
- Check for change in pumpage

Spring hang-up

- Improve environmental controls
- Change to seal with protected springs

Secondary hang-up

- Change to non-fretting mechanical seal

Incorrect installation

- Skills training needed
- Change to cartridge seal design

Crystallization

- Improve environmental controls

Flashing at seal faces

- Improve environmental controls
- Check pump operation for higher temperature

High percentage of solids

- Improve environmental controls
- Change to large bore seal chamber
- Check for change in pumpage

Repetitive/cyclical failure

- Check for low flow. (See Figure 11.1)
- Increase the shaft strength. (See 11.4.2.2)
- Improve environmental controls.
- Check inlet piping layout (See 11.4.2.1 and 9.2)
- Eliminate pipe strain
- Improve shaft alignment

Vibration problems

- See Figures 11.2 and 11.3

Excessive shaft end play

- See Bearings, Chapter 11.5.5

11.5.7.1 Mechanical seal – seal faces

Widened wear track

- See Bearings, Chapter 11.5.5
- Correct angularity of stationary face
- Increase the shaft strength. (See 11.4.2.2)
- Eliminate pipe strain

Zero wear track

- Improve seal installation to contact faces
- Change to large bore seal chamber

Severe uniform wear

- See Bearings, Chapter 11.5.5
- Check for excessive Pressure-Velocity value
- Improve environmental controls

Intermittent wear pattern

- Change to flexible mounted stationary face
- Ensure faces are flat when installed
- Installation training needed

Edge chipping

- Check pump operation for vaporization
- Improve environmental controls
- Increase the shaft strength. (See 11.4.2.2)

Heating checking

- Check for pumpage for excessive temperatures
- Check for excessive Pressure-Velocity value
- Check for dry running of pump
- Upgrade the material selection
- Improve environmental controls

Coking or Crystallization

- Check for pumpage for excessive temperatures
- Change to large bore seal chamber
- Improve environmental controls
- Upgrade the material selection

Blistering

- Upgrade the material selection

Corrosion

- Upgrade the material selection

Cracking or Fracture

- Check for rapid temperature fluctuations
- Check for excessive Pressure-Velocity value
- Installation/handling training needed
- Upgrade the material selection.

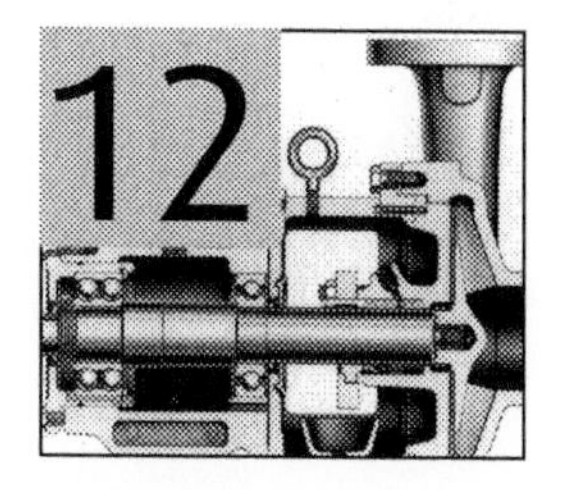

Pump maintenance

12.1 The strategy

At one time, the ultimate goal for a pump repair was to bring the unit back to the 'as new' standard of the original equipment manufacturer (OEM). Combining this with a corporate strategy of pump repair, there was a tendency to change out the part that failed and get the pump back into service as quickly as possible. However, if we establish that the current design is not able to meet all the various demands of the system, then a simple repair is not sufficient, and an upgrade becomes necessary. When such is the case, the necessary time must be allocated and scheduled for such work. If a failure has already taken place, and production is down, it may be necessary to schedule the upgrade at some time in the future and conduct a temporary repair immediately.

More enlightened companies operate with a strategy of Pump Reliability where every pump downtime event becomes an opportunity to upgrade the pump to a higher level of reliability. With this in mind, a Pump Reliability strategy consists of continuous equipment monitoring, detection of developing problems, in-depth diagnosis and corrective action.

To conduct an effective upgrading of any pump, it is necessary to ensure that all the information needed is available to the personnel involved. This should include:

1. Access to all prior repair records.
2. The installation, operation and maintenance manual with data sheet, bills of materials, the pump curve and the sectional drawing.
3. Mechanical seal bill of materials and installation drawing.
4. All necessary repair fits and tolerances.

5. Practical knowledge of pump design.
6. Component inspection equipment and techniques.
7. Sound repair techniques.
8. Practical machining skills.
9. Access to a well equipped and clean repair facility with quality machining and balancing equipment.
10. Established acceptance criteria.

12.1.2 Economic considerations

It is important to remember, before any repair procedures are performed on any pump component, that the material of construction must be accurately identified by means of the appropriate tests. It is also advisable to consider the economic advantage of the repair under consideration.

Smaller and medium-sized ANSI pumps are designed with a high degree of interchangeability and produced in volume. Consequently, it can frequently be more cost effective to replace the entire pump rather than a combination of the impeller, casing and back cover. In addition, both the individual parts and complete pumps are available fairly quickly. This can make it more cost effective to replace rather than repair the parts, unless the wet ends are made of the more exotic alloys. It is clear, in the case of non-metallic pumps, that the components must be replaced, as they generally cannot be repaired.

API pumps, however, are generally more economical to repair than to replace. These units are usually installed in more rugged duties and hazardous applications in refineries or other petrochemical industries, and are consequently more durable and more expensive. Delivery periods are also frequently longer, and the parts more costly than their ANSI equivalents, particularly the cases and impellers.

This makes it very tempting to source these parts from an after-market supplier rather than the Original Equipment Manufacturer (OEM). However, it should be noted that the major parts of a centrifugal pump (i.e. the casing, the impeller and the back cover) are all cast from patterns involving highly engineered hydraulic designs, which are of a proprietary nature. These parts are also the ones that provide the hydraulic performance of the pump. While the parts might be available from after-market suppliers at slightly lower prices than they are from the OEM, that cost saving will fade into insignificance if the pump does not meet its hydraulic performance. The OEM can accept the responsibility for the subsequent hydraulic performance of these replacement parts.

An additional challenge in the financial aspect is the ability to persuade senior management that the additional expenditure sometimes required for a Reliability approach has quantifiable benefits. As most senior managers are not always aware of the technicalities involved in pump repair, or the ramifications of such repair, it becomes the responsibility of the maintenance personnel to effect the translation from the technicalities to the financial benefits. The ability to translate a prevented failure into dollars saved is critical.

The most important number that must be learned is the cost of down-time for the equipment (or system). This could involve the loss of sales revenue from the product being produced and the loss of productivity of the personnel involved. As this number is frequently in the tens of thousands of dollars per hour of lost production, it become a critical factor in the (financial) decision making process.

It is also important to know the difference in the cost of a repair conducted in a planned and organized fashion as opposed to one conducted in a reactive mode. The latter usually includes a lot of time wasted while personnel, tools and spares are sourced and brought to site. With a planned outage, all necessary personnel and equipment are scheduled to be on site at the time the job is due to begin, thus minimizing the risk of wastage. The difference in costs of such a repair has been reported to be anywhere from 25% to 50% of the cost of an unplanned reactive repair.

For Reliability strategy to be a success it is essential that support be gained from those holding the purse strings. Consequently the translation from technical benefits to dollar figures for each strategy proposed must be undertaken by the maintenance staff in order secure such support.

12.2 Preparation for pump dismantling

To ensure that each component of the pump is properly inspected, it is recommended that standardized dismantling techniques be implemented. This will ensure that every pump of the same type is dismantled in the same way, with every step systematized and all measurements recorded in the same way, for reference.

The first order of business in dismantling a pump for service is to lock out the electrical power in order to prevent accidental startup of the driving motor, and physical injury. In addition, all isolating valves on both the pump suction and discharge side must be shut off.

If the pump has been operating at high temperature, allow sufficient time for it to cool down before trying to dismantling the equipment. At this point, all joints and piping should be match marked for accurate reassembly.

It is also important to remove any hazardous liquids and vapors from all parts of the pump before bringing that unit into a confined area such as the repair facility. This is particularly necessary if the pump is being shipped out to an off-site location for repair, as that outside facility may not be aware of the manner in which a particular hazardous material must be handled. With any hazardous liquids a material safety data sheet on that liquid should accompany the pump whenever it leaves the plant.

While a decontamination process may be an obvious step for the wet end assembly, it is not always so apparent for the bearing housing even although the possibility always exists that excessive leakage from a stuffing box could allow contaminated process liquid to enter the bearing housing.

If your company conducts a regular lubricating oil analysis, the oil should be drained from the bearing housing and a sample retained in a clean container for inspection. In some cases that oil sample may assist in identifying the cause of component failure. Once the oil is drained, the bearing housing should be flushed and drained several times with an acceptable solvent.

During dismantling, all wear parts should be collected, marked and stored for inspection. This information will be useful for confirming or identifying the root cause of the problem and determining the necessary upgrades.

12.3 Removing the back pull-out assembly

To remove the back pull-out assembly of the end suction centrifugal pump, place the sling through the pump frame adapter and have the crane or hoist take the strain before removing the bearing housing hold-down bolts and the casing/adapter bolts. The casing jacking screws can then be tightened to withdraw the back pull-out assembly from the pump casing.

At that point, the assembly can be removed to a clean work bench in a clean environment. To facilitate further dismantling of the power end, it is recommended that the bearing housing be secured to the workbench by the bearing housing foot, and the adapter supported until removed.

12.4 Inspection checks on cast parts

As the pump comes apart, the overall condition of the parts should be visually inspected for any signs of cracks, surface damage or other evidence of distress. Particularly take note of erosion, corrosion or wear patterns.

The pump casing should be examined in the impeller running surface area, the wear ring area, the volute, the gasket surface and register fit. If localized wear, pitting or grooving greater than 0.125 inches. deep is evident, the casing should be replaced or repaired. All alignment and gasket fits should be inspected for indications of surface damage.

With an open impeller pump, the wear pattern on the contoured surface of the casing on which the impeller matches should be considered for indications of shaft deflection or casing misalignment.

In a closed impeller pump, inspect the wear ring locking mechanism for cracks or other signs of distress. The wear ring(s) should be replaced if the clearance is greater than 150% of design. Consult the pump supplier for the wear ring design clearance in that particular service.

The impeller must be inspected for excessive wear, pitting and impact damage. This is particularly the case on the edges of the vanes of an open impeller. Any grooving deeper than 0.060 inches, or an even wear of more than 0.030 inches should require replacement or repair.

Once the visual inspection is complete, it may be appropriate to subject the casing to non-destruction testing (NTD) procedures to identify any cracks.

In the event that any weld repair is needed on the casing, it must be conducted in accordance with the necessary standard and by qualified personnel, in order to be classified as a long term repair.

When a casing has been damaged by erosion, the casing can be repaired by applying a coating to reclaim the worn areas. As many aspects of the coating and it's compatibility to the material are critical in the successful completion of such a repair, it is recommended that this not be considered a do-it-yourself job, and a qualified contractor be used. The chemical compatibility of the coating must also be carefully checked for compatibility with the process fluid. While the same procedures can be used against corrosion, it is usually only considered a short term repair.

12.5 Casing and wear rings

The following recommendations and procedures are offered as general guidelines only for the wet end of the pump. When available, the pump manufacturer's operation and maintenance instructions are usually the best source of information for pump assembly.

1. Check the casing register fit and gasket surface for concentricity, parallelism and perpendicularity.
2. Skim the wear ring fit by 0.001 to 0.002 inches. to ensure concentricity with the registered fits.
3. Machine a new wear ring with an OD of 0.002 to 0.003 inches. larger than the casing bore.
4. Install the wear ring in the housing and lock it into place by either tack welding in 3 or 4 locations, or by set screws. Once the set screws are secured, peen the ring to prevent them from backing out.
5. Machine the I.D. of the wear ring to ensure the necessary clearance for the impeller in accordance with the manufacturer's instructions.
6. Check the casing register to ensure the correct clearance of 0.0004 inches. or less between the casing register fit and the back cover or frame adapter.
7. Check to ensure that the register fit is parallel to the register fit centerline within 0.002 inches, or perpendicular to within 0.001 inches, as appropriate.
8. The gasket surface also should be cleaned up to have a maximum of 0.003 inches of runout and be free from any pitting or scratches in the sealing surface. In addition, it should be perpendicular to the register fit centerline within 0.002 inches.

While a closed impeller pump is being reassembled, check the radial clearance between the impeller and casing wear rings as this will provide a good indicator of the assembled alignment between the mating parts. The radial clearance should be equal to or within 0.001 inch around the diameter.

12.6 Dismantling the back pull-out assembly

As the assembly is being dismantled, any shims under the bearing housing foot should be collected, marked and stored for inspection. This also applies to the coupling hub and key.

When attempting to remove an impeller, it is important never to apply heat as it may cause an explosion as a result of trapped fluid, possibly causing severe personal injury and property damage. In addition it is recommended that heavy work gloves be used when handling an open impeller as the vane edges may be sharp enough to cut.

12.6.1 When packing is fitted

Withdraw the gland and remove the back cover and stuffing box from the adapter. Remove the shaft sleeve and gland from the shaft, and then remove the packing and lantern ring from the stuffing box.

12.6.2 When a mechanical seal is fitted

Draw the gland back carefully, bearing in mind that the stationary face of the mechanical seal is positioned in the gland.

Carefully remove the stuffing box or seal chamber from over the mechanical seal rotating element which is mounted on the sleeve.

Remove the rotating element of the seal from the shaft sleeve. Then remove the gland from the shaft, and the stationary face of the seal from the gland.

At this point it is now possible to remove the dowel pins and unbolt the frame adapter from the bearing housing.

The gasket between the adapter and bearing housing should now be removed and the gasket surfaces cleaned for a new gasket to be installed on reassembly.

The flinger and the lip seal or bearing isolator closest to the impeller end of the shaft can now be removed.

12.6.3 Dismantling the bearing housing

1. Remove the bearing cap securing bolts.
2. To remove the shaft assembly, tighten the jack screws evenly until the bearing cap clears the housing, at which time the shaft assembly can be drawn clear of the bearing housing.
3. Remove the bearing cap 'O'-ring.
4. Remove the snap ring that locates the thrust bearing in the cap and then withdraw the bearing cap from the shaft, leaving the bearings in position.
5. Remove the lip seal or the bearing isolator from the bearing cap.
6. Withdraw the radial bearing from shaft.
7. Remove the thrust bearing locknut and lockwasher.
8. Withdraw the thrust bearing from shaft.

When removing a bearing from the shaft, it is important to use force on the inner ring of the bearing only. Save the bearings for inspection.

12.7 Inspection checks

12.7.1 The back cover

The integral parts of the back cover and stuffing box (or seal chamber) should be inspected for all pitting and wear damage as well as for cracks or excessive corrosion or erosion damage. If localized wear, pitting or grooving is greater than 0.125 inches deep, the cover should be replaced or repaired. Any irregularities in the gasket seating surfaces should be cleaned up.

If frequent seal failures are being experienced, a sound Reliability Strategy would consider replacing a traditional stuffing box with a large bore seal chamber as they have been proved to assist in the reliability of mechanical seals in process pumps.

12.7.2 The frame adapter

This should be inspected for cracks and evidence of excessive corrosion damage.

12.7.3 The bearing housing

The bearing housing should also be inspected for cracks and evidence of excessive corrosion damage. The inside surfaces of the housing should be cleared of all loose material such as rust, scale and other debris. All lubrication passages should be cleared. In addition, the radial fit on the bore of the radial bearing must also be checked to ensure that it is still within the appropriate tolerances. This also applies to the bore fit of the thrust bearing in the bearing cap.

A widely used upgrade is to replace the constant level oiler with a bullseye type of oil sight glass. An oil mist system as discussed in Chapter 7 would be considered a sound Reliability Strategy.

12.7.4 Lip seals

If a lip seal is removed, a sound Reliability Strategy will require that it, and the flinger, be discarded and – when the pump is reassembled – they be replaced with a suitable bearing isolator. If the isolator is already in place, it can be stored for reinstallation when the pump is reassembled.

12.7.5 The shaft

Both the shaft and the sleeve (if fitted) must be inspected for grooves and pitting damage. All dimensions must be within tolerances, especially the bearing fits. The shaft must also have a run-out within the usual tolerance of 0.002 inches.

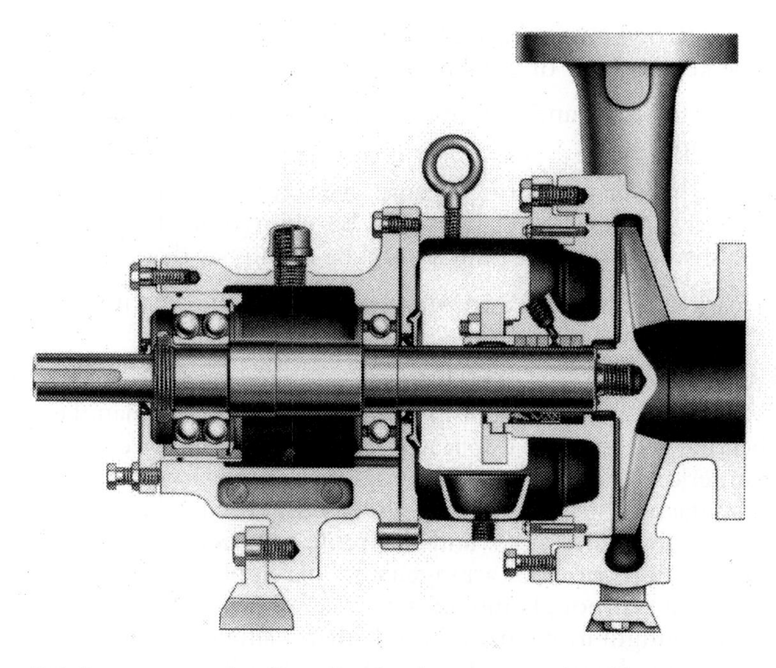

Figure 12.1: Pump cross section (Reproduced with permission of Goulds Pumps, ITT Industries)

If a sleeve has been used, a determination should be made at this point if that practice will continue. With the many mechanical seals on the market that do not impose fretting damage on the shaft, the protective sleeve is no longer required. In addition to which, a larger diameter shaft (to the same O.D. of the sleeve) provides a higher resistance to deflection and bending which increases the reliability of seals and bearings.

This opportunity to fit a stronger shaft, together with the other upgrades identified in this Chapter, is in line with the consistent upgrading of all the equipment in the plant, and contributes to a Total Reliable Strategy.

12.8 Mounting bearings on the shaft

Please take note that the following information is intended only as a general guideline, and should not be considered as detailed instruction. The most reliable and up-to-date source of information will be the manufacturers' installation, operation and maintenance manual.

For a bearing to function properly, it is essential that the utmost care be taken with regard to cleanliness. Wherever possible, bearing installation should be carried out in a clean and dust-free room. New bearings should be kept in their original package until just before installation.

1. Clean the shaft and the mounting surfaces of the bearing housing and cap.
2. Heat the Thrust bearing in an induction heater or oil bath to the required temperature. This will be approximately 150°F above the shaft temperature, but should be no higher than 250°F.
3. Wearing clean protective gloves, position the bearing onto the shaft against the shoulder.

   When installing a new bearing into a pump, it is extremely important that the pump original equipment manufacturer's bearing selection be duplicated during repair and overhaul. While numerical suffixes can be cross-referenced from one bearing manufacturer to another, the alphabetical prefixes and suffixes are frequently different, and extreme caution is encouraged. Duplication of the arrangement of duplex bearings is also important although most pump manufacturers use the back-to-back configuration. In the event that a single shield is provided, this should be on the side towards the impeller.
4. Position the lockwasher on the shaft with the tang in the shaft keyway.
5. Thread the locknut onto the shaft and tighten until snug. Bend any tang of the lockwasher into a slot of the locknut.
6. Position the thrust bearing retaining ring over the shaft with the flat side facing the bearing.
7. Heat the Radial bearing in an induction heater or oil bath to the required temperature. This will be approximately 150°F above the shaft temperature, but should be no higher than 250°F.
8. Wearing clean protective gloves, position the bearing onto the shaft against the shoulder.
9. Coat the outside of the thrust bearing and bearing cap with oil and install the bearing cap onto the shaft/bearing assembly.
10. Insert the retaining ring into the groove in the bearing cap bore, while ensuring that the space between the ends of the retaining ring is located in the oil return groove so as not to obstruct the oil flow.
11. Check to ensure that the shaft is free to turn.
12. Ensure the keyway is free of burrs. Cover it with tape to protect the 'O' rings on the bearing isolator.

13. Install the outboard bearing isolator and ensure that the drain slot is located at the bottom (or 6 o'clock) position.
14. Install a new 'O' ring on the outside of the bearing cap and coat the O.D. with oil.
15. Coat the internal surfaces of the bearing housing with oil and install the shaft assembly into the housing.
16. Once again, check to ensure that the shaft is free to turn.
17. Tighten the bolt fasteners through the bearing cap to the housing and hand tighten.
18. Install the jacking bolts with locking nuts and hand tighten.

12.8.1 Check shaft condition

Support the bearing housing assembly in the horizontal position and check the following conditions.

1. Confirm that the shaft end play falls within the limitations established for the particular size of pump and type of bearing.
2. With the impeller threaded onto the shaft, confirm that the shaft run-out, when the shaft is rotated through 360°, is no more than a total indicator reading of 0.002 inches.
3. By rotating the shaft so that indicator rides along the housing face fit throughout the 360°, ensure that the total indicator reading is no more than 0.001 inches.

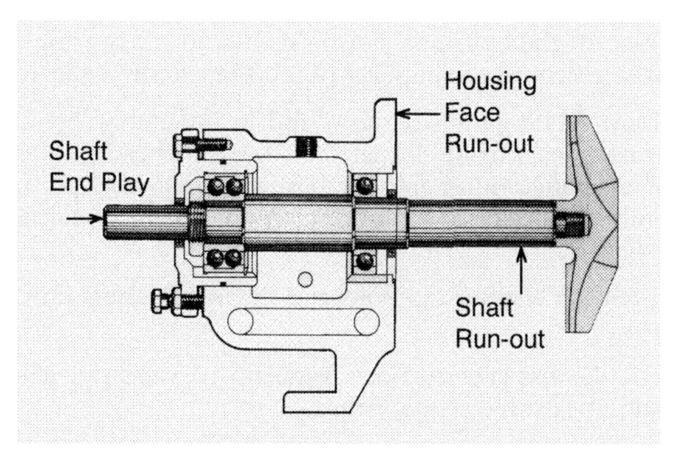

Figure 12.2: Shaft run-out and end play

12.9 Assembling the back pull-out assembly

1. Install the frame adapter to the bearing housing with the gasket.
2. Tighten bolts in accordance with the torque specifications of the manufacturer in a criss-cross pattern.
3. Check adapter fits and ensure the total indicator reading is no greater than 0.005 inches by rotating the shaft through 360°
4. Install the bearing isolator into the adapter and ensure that the drain slot is located at the bottom (or 6 o'clock) position.

12.9.1 A packed stuffing box

1. Install the back cover with the stuffing box
2. Check the back cover run-out by rotating the indicator through 360°. The total indicator reading should be within 0.005 inches.
3. Install the shaft sleeve until it is fully seated, with anti-galling compound to facilitate the next disassembly.

12.9.1.1 Installing the impeller with a packed stuffing box

1. Using heavy work gloves to protect against the sharp edges of the impeller, install the impeller with the 'O' ring seal until the impeller makes firm contact with the shaft.
2. Tighten the impeller onto the shaft in accordance with the pump manufacturer's instructions.
3. Loosen the bolts and jacking screws on the bearing cap.
4. With feeler gauges, measure the gap between the rear of the impeller and the face of the back cover.
5. Loosen off the bolts and jacking screws on the bearing cap until a clearance behind the impeller of 0.030 inches is achieved.

   In most ANSI pump designs, and in an as-new condition, this approximates a front clearance of 0.015 inches from casing. This final clearance must be established when the pump is fully assembled, just prior to locking the mechanical seal down onto the shaft.
6. Tighten the bearing cap bolts and the locking nuts on the jacking screws.
7. Check the impeller run-out, vane tip to vane tip. The run-out should not exceed 0.005 inches.

12.9.1.2 Packing procedures

1. Ensure that all the old packing has been cleaned out of the stuffing box, and clean the stuffing box and the sleeve thoroughly. The sleeve should be replaced if any significant wear is identified.
2. Measure the bore of the stuffing box and the diameter of the shaft. Subtract the O.D. dimension from the I.D. dimension and divide by 2. The result is the required size of the packing.
3. When using coil or spiral packing, always cut the packing to size. Do not wind the packing into the stuffing box around the shaft. When cutting the packing, always use a mandrel of the same size as the shaft in the stuffing box.
4. Hold the packing lightly on the mandrel, but do not stretch it. Cut all the rings in the same manner, either with a Butt Joint or with a Skive Joint as shown.
5. Owing to the importance of cutting the rings to the correct size, Die-cut rings can be of great advantage as they give you the exact size ring for the I.D. of the box and the O.D. of the shaft.

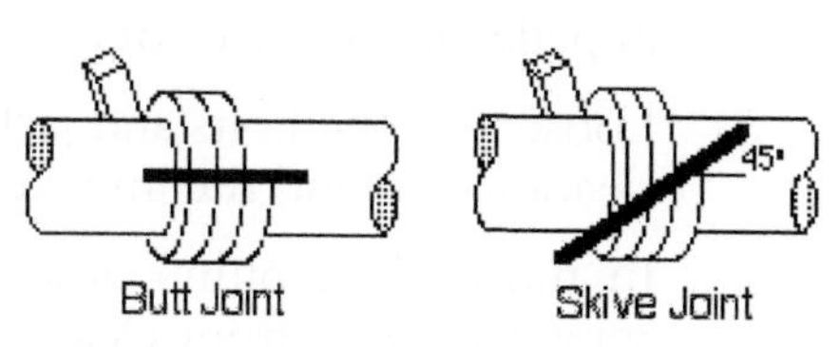

Figure 12.3: Cutting packing rings

6. Install one ring at a time and make sure each ring stays clean prior to installation.
7. Seat rings snugly against each other and the joints of successive rings should be staggered and kept at least 90° apart.
8. After the last ring is installed, take up the gland bolts finger tight only. Do not jam the packing into place by excessive gland loading.
9. Start pump, and allow the packing to leak freely on a new pump or one that has just been overhauled. Excessive leakage during the first hour of operation will result in more reliable packing arrangement over a long period of time.
10. Gradually tighten up on the gland as the packing settles, until the leakage is reduced to an acceptable level in accordance with plant standards.

Note that all packing is permitted some slight amount of leakage in order to provide cooling and lubrication between the packing and the sleeve.

12.9.2 Installing with a mechanical seal

1. Install the back cover with the seal chamber.
2. Check the back cover run-out by rotating the indicator through 360° Total indicator reading should be within 0.005 inches.
3. Install the shaft sleeve until it is fully seated.

12.9.2.1 Installing the impeller with a mechanical seal

1. Using heavy work gloves to protect against the sharp edges of the impeller install the impeller with the 'O' ring sea until the impeller makes firm contact with the shaft.
2. Tighten the impeller onto the shaft in accordance with the pump manufacturer's instructions.
3. Loosen the bolts and jacking screws on the bearing cap.
4. With feeler gauges, measure the gap between the rear of the impeller and the face of the back cover.
5. Loosen off the bolts and jacking screws on the bearing cap until a clearance behind the impeller of 0.030 inches is achieved.

   In most ANSI pump designs, and in an as-new condition, this approximates a front clearance of 0.015 inches from casing. This final clearance must be established when the pump is fully assembled, just prior to locking the mechanical seal down onto the shaft.
6. Tighten the bearing cap bolts and the locking nuts on the jacking screws.
7. Check the impeller run-out, vane tip to vane tip. The run-out should not exceed 0.005 inches.
8. Scribe a mark on the shaft at the position of the face of the seal chamber. This will be the datum for the installation of the mechanical seal.
9. Remove the impeller and seal chamber.
10. Obtain the Seal Installation Reference measurement from the seal manufacturer's drawing. This is the distance from the face of the stuffing box to the rear location of the mechanical seal. Inscribe that point on the shaft.
11. Position the stationary face into the gland and slide them over the shaft towards the bearings.
12. Install the mechanical seal on the shaft in accordance with the seal manufacturer's instructions.

13. Reinstall the impeller as described in (2.) above and recheck run-out.

12.10 Installing the back pull-out assembly

1. Clean the casing, fit and install the casing gasket on the back cover.
2. Loosen the bolts and jacking screws on the thrust bearing cap.
3. Install the back pull-out assembly in the casing.
4. Install the casing bolts finger tight, then tighten them to the torque value required by the manufacturer.
5. Install the casing jacking screws as a snug fit. Do not overtighten.
6. Ensure that the total travel of the impeller in the casing is within the pump manufacturer's tolerances. Most ANSI pump designs are between 0.030 inches and 0.065 inches.
7. Tighten the thrust bearing cap bolts evenly, drawing the cap towards the bearing housing, until the impeller just contacts the casing.
8. Set the dial indicator so that the button contacts the end of the shaft.
9. Set the dial indicator to zero and back off the bolts about one turn.
10. Tighten the jacking screws evenly (about one flat at a time), thus backing away the bearing cap from the bearing housing until the dial indicator shows the required clearance.
11. Evenly tighten the bolts and jacking screws, keeping the indicator at the required setting.
12. Check the shaft to ensure it still turns freely.

Whenever possible, the same people who start any repair or upgrade on a particular pump should see that job all the way through. This will minimize a potential problem arising from miscommunication regarding what has and what has not been done during the repair procedure.

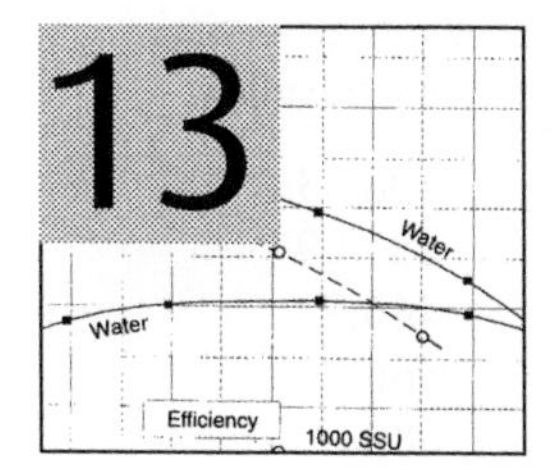

Fluid properties

13.1 Properties of water at various temperatures

Table 13.1: Properties of water at various temperatures

Temperature °F	Temperature °C	Specific gravity 60 F reference	Vapor pressure psi absolute
32	0	1.001	0.088
40	4.4	1.001	0.1217
50	10.0	1.001	0.1781
60	15.6	1.000	0.2563
70	21.1	0.999	0.3631
80	26.7	0.998	0.5069
90	32.2	0.996	0.6982
100	37.8	0.994	0.9492
120	48.9	0.990	1.692
140	60.0	0.985	2.889
160	71.1	0.979	4.741
180	82.2	0.972	7.510
200	93.3	0.964	11.526
212	100.0	0.959	14.696
220	104.4	0.956	17.186
240	115.6	0.948	24.97
260	126.7	0.939	35.43
280	137.8	0.929	49.20
300	148.9	0.919	67.01
320	160.0	0.909	89.66
340	171.1	0.898	118.01
360	182.2	0.886	153.04
380	193.3	0.874	195.77
400	204.4	0.860	247.31

Table 13.1: continued

Temperature °F	Temperature °C	Specific gravity 60 F reference	Vapor pressure psi absolute
420	215.6	0.847	308.83
440	226.7	0.833	381.59
460	237.8	0.818	466.9
480	248.9	0.802	566.1
500	260.0	0.786	680.8
520	271.1	0.766	812.4
540	282.2	0.747	962.5
560	293.3	0.727	1133.1
580	304.4	0.704	1325.8
600	315.6	0.679	1542.9
620	326.7	0.650	1786.6
640	337.8	0.618	2059.7
660	348.9	0.577	2365.4
680	360.0	0.526	2708.1
700	371.1	0.435	3093.7
705.4	374.1	0.319	3206.2

13.2 Effect of altitude on pressures and the boiling point of water

Table 13.2: Effect of altitude on pressures and the boiling point of water

Altitude		Barometric reading		Atmospheric pressure			Boiling point of water	
Feet	Meters	ins. Hg.	mm. Hg	psia	Feet of water	Kg.cm2	°F	°C
0	0.0	29.9	760	14.7	33.9	1.033	212.0	100.0
500	152.4	29.4	747	14.4	33.3	1.015	211.1	99.5
1000	304.8	28.9	734	14.2	32.8	1.000	210.2	99.0
2000	609.6	27.8	706	13.7	31.5	0.960	208.4	98.0
3000	914.4	26.8	681	13.2	30.4	0.927	206.5	96.9
4000	1219.2	25.8	655	12.7	29.2	0.890	204.7	95.9
5000	1524.0	24.9	633	12.2	28.2	0.860	202.9	94.9
6000	1828.8	24.0	610	11.8	27.2	0.829	201.0	93.9
7000	2133.6	23.1	587	11.3	26.2	0.799	199.2	92.9
8000	2438.4	22.2	564	10.9	25.2	0.768	197.4	91.9
9000	2743.2	21.4	544	10.5	24.3	0.741	195.5	90.8
10000	3048.0	20.6	523	10.1	23.4	0.713	193.7	89.8
15000	4572.0	16.9	429	8.3	19.2	0.585	184.0	84.4

13.3 Viscous liquids

When handling viscous liquids with a centrifugal pump, some marked differences are noticed from the values that are identified on the characteristic performance curves. There is a marked increase in the horsepower draw, a reduction in total head and some reduction in capacity.

The charts under Figures 13.3 and 13.4 have been designed to assist with the calculation of the viscous performance of a standard centrifugal pump when the regular water performance is known.

As an example, consider the conditions where the pump is required to deliver 750 gpm at a total head of 100 feet, when the liquid has a viscosity of 1,000 SSU and a specific gravity of 0.9 at the pumping temperature.

Enter Figure 13.4 at the bottom at the flow of 750 gpm, go up to a Head of 100 feet. From that point, move horizontally on the chart to the viscosity of 1,000 SSU, and then vertical upwards to intersect the curves for the correction factors.

The lowest curve C_h provides the correction for the Efficiency at 0.635

The next curve C_Q identifies the correction for Capacity at 0.95

The next series of curves C_H identifies the correction for the Total Head at various percentages of the B.E.P., and at the $1.0 \times Q_{NW}$ (or B.E.P) the correction is 0.92

$$Qw = \frac{750}{0.95} = 790 \text{ gpm}$$

$$Hw = \frac{100}{0.92} = 109 \text{ ft. of Head}$$

With this data, we can now select a pump with a water capacity of 790 gpm at a head of 109 feet. The selection should be made as close as possible to the maximum efficiency point (B.E.P.) for water performance. At that condition on the sample performance Figure 13.6 the efficiency is identified as 81%. Consequently the efficiency for the viscous liquid will be as shown in the formula below.

$$E_{vis.} = 0.635 \times 81\% = 51.5\%$$

The brake horsepower required for pumping the viscous liquid will be as shown below

$$BHP_{vis.} = \frac{750 \times 100 \times 0.9}{3960 \times 0.515} = 33.1 \text{ bhp}$$

To show the viscous performance over a wider range, the same calculations can be made at the other values of C_H. Thus providing the kind of information shown on Figure 13.5, which can then be plotted on the characteristic pump curve showing the variation in performance as on 13.6.

13.3.1 Viscous performance correction for small pumps

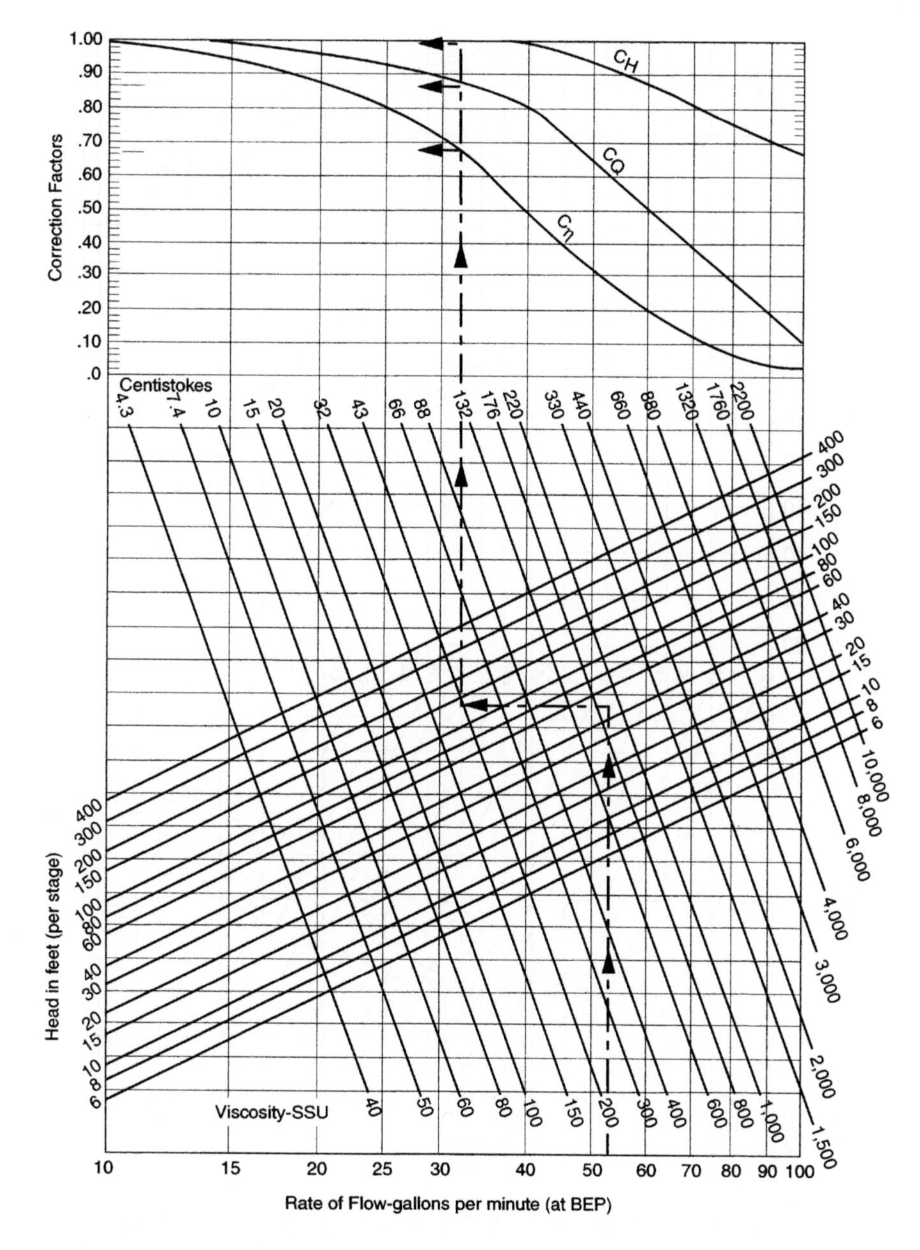

Figure 13.3: Performance correction chart for small pumps (Reprinted by permission of the Hydraulic Institute)

13.3.2 Viscous performance correction for larger pumps (over 100 gpm)

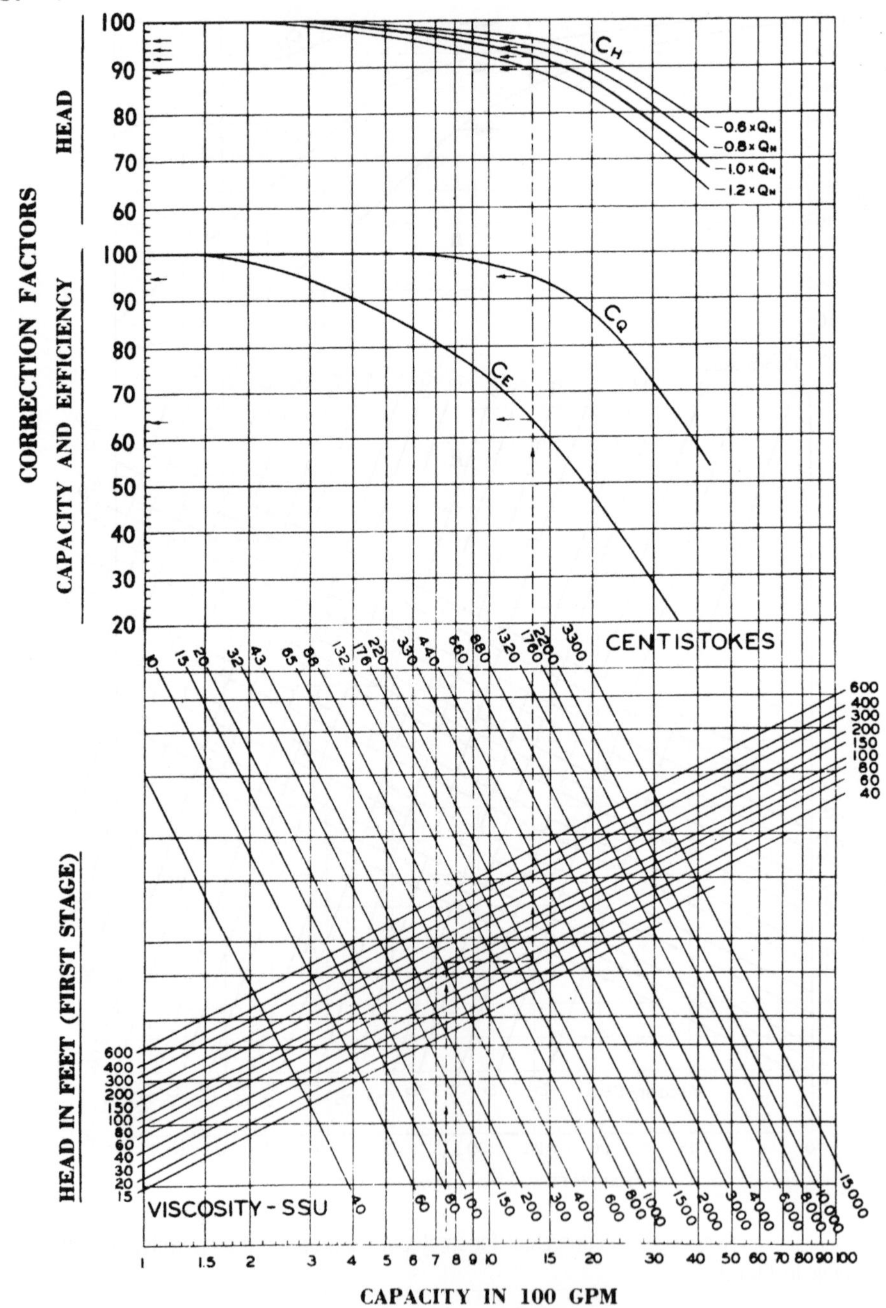

Figure 13.4: Performance correction chart for large pumps (Reprinted with permission of the Hydraulic Institute)

	0.6 x Q_{nw}	0.8 x Q_{nw}	1.0 x Q_{nw}	1.2 x Q_{nw}
Water rate of flow (Q_W) - gpm	450	600	750	900
Water head in feet (H_W) - feet	114	108	100	86
Water efficiency (η_W) - %	72.5	80	82	79.5
Viscosity of liquid - SSU	1000	1000	1000	1000
C_Q from chart	0.95	0.95	0.95	0.95
C_H from chart	0.96	0.94	0.92	0.89
C_η from chart	0.635	0.635	0.635	0.635
Viscous rate of flow: $Q_W \times C_Q$ - gpm	427	570	712	855
Viscous head: $H_W \times C_H$ - feet	109.5	101.5	92	76.5
Viscous efficiency: $\eta_W \times C_\eta$ - %	46.0	50.8	52.1	50.5
Specific gravity of liquid	0.90	0.90	0.90	0.90
Viscous power - hp	23.1	25.9	28.6	29.4

Figure 13.5: Sample calculations (Reprinted with permission of Hydraulic Institute)

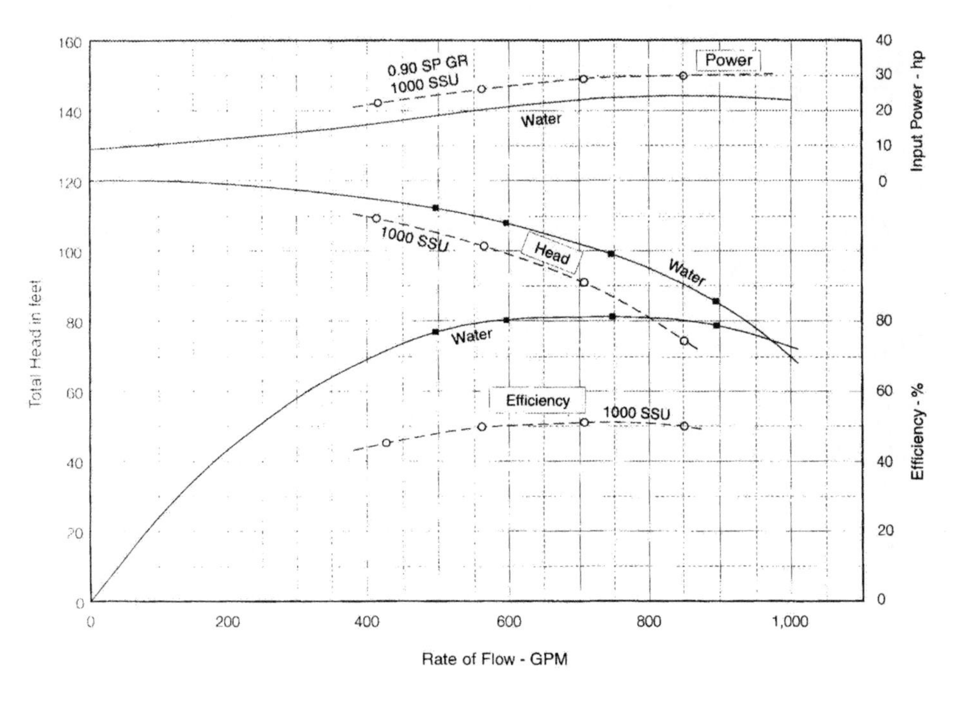

Figure 13.6: Sample performance chart (Reprinted with permission of the Hydraulic Institute)

Seconds Saybolt Universal ssu	Kinematic Viscosity Centistokes *	Seconds Saybolt Furol ssf	Seconds Redwood 1 (Standard)	Seconds Redwood 2 (Admiralty)	Degrees Engler	Degrees Barbey	Seconds Parlin Cup #7	Seconds Parlin Cup #10	Seconds Parlin Cup #15	Seconds Parlin Cup #20	Seconds Ford Cup #3	Seconds Ford Cup #4
31	1.00	—	29	—	1.00	6200	—	—	—	—	—	—
35	2.56	—	32.1	—	1.16	2420	—	—	—	—	—	—
40	4.30	—	36.2	5.10	1.31	1440	—	—	—	—	—	—
50	7.40	—	44.3	5.83	1.58	838	—	—	—	—	—	—
60	10.3	—	52.3	6.77	1.88	618	—	—	—	—	—	—
70	13.1	12.95	60.9	7.60	2.17	483	—	—	—	—	—	—
80	15.7	13.70	69.2	8.44	2.45	404	—	—	—	—	—	—
90	18.2	14.44	77.6	9.30	2.73	348	—	—	—	—	—	—
100	20.6	15.24	85.6	10.12	3.02	307	—	—	—	—	—	—
150	32.1	19.30	128	14.48	4.48	195	—	—	—	—	—	—
200	43.2	23.5	170	18.90	5.92	144	40	—	—	—	—	—
250	54.0	28.0	212	23.45	7.35	114	46	—	—	—	—	—
300	65.0	32.5	254	28.0	8.79	95	52.5	15	6.0	3.0	30	20
400	87.60	41.9	338	37.1	11.70	70.8	66	21	7.2	3.2	42	28
500	110.0	51.6	423	46.2	14.60	56.4	79	25	7.8	3.4	50	34
600	132	61.4	508	55.4	17.50	47.0	92	30	8.5	3.6	58	40
700	154	71.1	592	64.6	20.45	40.3	106	35	9.0	3.9	67	45
800	176	81.0	677	73.8	23.35	35.2	120	39	9.8	4.1	74	50
900	198	91.0	762	83.0	26.30	31.3	135	41	10.7	4.3	82	57
1000	220	100.7	896	92.1	29.20	28.2	149	43	11.5	4.5	90	62
1500	330	150	1270	138.2	43.80	18.7	—	65	15.2	6.3	132	90
2000	440	200	1690	184.2	58.40	14.1	—	86	19.5	7.5	172	118
2500	550	250	2120	230	73.0	11.3	—	108	24	9	218	147
3000	660	300	2540	276	87.60	9.4	—	129	28.5	11	258	172
4000	880	400	3380	368	117.0	7.05	—	172	37	14	337	230
5000	1100	500	4230	461	146	5.64	—	215	47	18	425	290
6000	1320	600	5080	553	175	4.70	—	258	57	22	520	350
7000	1540	700	5920	645	204.5	4.03	—	300	67	25	600	410
8000	1760	800	6770	737	233.5	3.52	—	344	76	29	680	465
9000	1980	900	7620	829	263	3.13	—	387	86	32	780	520
10000	2200	1000	8460	921	292	2.82	—	430	96	35	850	575
15000	3300	1500	13700	—	438	2.50	—	650	147	53	1280	860
20000	4400	2000	18400	—	584	1.40	—	860	203	70	1715	1150

Figure 13.7: Viscosity conversion table (Reprinted with permission of the Hydraulic Institute)

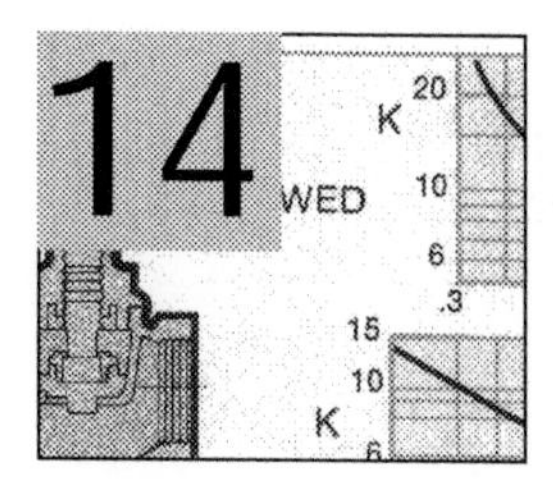

Friction loss tables

14.1 Friction loss for water in pipe

1 INCH NOMINAL		STEEL SCHEDULE 40 ID = 1.049 INCHES ϵ/D = 0.00172		
DISCHARGE		V	$V^2/2g$	h_f
CFS	GPM	ft/sec	feet	feet per 100 feet of pipe
0.00223	1	0.371	0.00214	0.114
0.00446	2	0.742	0.00857	0.379
0.0068	3	1.114	0.01927	0.772
0.00891	4	1.48	0.0343	1.295
0.0111	5	1.86	0.0535	1.93
0.0134	6	2.23	0.0771	2.68
0.0156	7	2.60	0.1049	3.56
0.0178	8	2.97	0.137	4.54
0.0201	9	3.34	0.173	5.65
0.0223	10	3.71	0.214	6.86
0.0267	12	4.45	0.308	9.62
0.0312	14	5.20	0.420	12.8
0.0356	16	5.94	0.548	16.5
0.0401	18	6.68	0.694	20.6
0.0446	20	7.42	0.857	25.1
0.0490	22	8.17	1.036	30.2
0.0535	24	8.91	1.23	35.6
0.0579	26	9.65	1.45	41.6
0.0624	28	10.39	1.68	47.9
0.0668	30	11.1	1.93	54.6
0.0713	32	11.9	2.19	61.8
0.0758	34	12.6	2.48	69.4
0.0802	36	13.4	2.78	77.4
0.0847	38	14.1	3.09	86.0
0.0891	40	14.8	3.43	95.0
0.0936	42	15.6	3.78	104.5
0.0980	44	16.3	4.15	114
0.102	46	17.1	4.53	124
0.107	48	17.8	4.93	135
0.111	50	18.6	5.35	146
0.123	55	20.4	6.48	176
0.134	60	22.3	7.71	209
0.145	65	24.1	9.05	245
0.156	70	26.0	10.49	283
0.167	75	27.8	12.0	324

Figure 14.1: Friction loss for water in feet per 100 ft. of 1 inch pipe (Reproduced with the permission of The Hydraulic Institute)

Note: No allowance has been made for age, differences in diameter, or any abnormal condition of interior surface. Any factor of safety must be estimated from the local conditions and the requirements of each particular installation.

1½ INCH NOMINAL		STEEL SCHEDULE 40 ID = 1.610 INCHES ε/D = 0.00112		
DISCHARGE		V	$V^2/2g$	h_f
CFS	GPM	ft/sec	feet	feet per 100 feet of pipe
0.00334	1.5	0.236	0.000868	0.0300
0.00446	2	0.315	0.00154	0.0492
0.00668	3	0.473	0.00347	0.0988
0.00891	4	0.630	0.00618	0.164
0.0111	5	0.788	0.00965	0.242
0.0134	6	0.946	0.0139	0.333
0.0156	7	1.103	0.0189	0.439
0.0178	8	1.26	0.0247	0.558
0.0201	9	1.42	0.0313	0.689
0.0223	10	1.58	0.0386	0.829
0.0267	12	1.89	0.0556	1.16
0.0312	14	2.21	0.0756	1.53
0.0356	16	2.52	0.0988	1.96
0.0401	18	2.84	0.125	2.42
0.0446	20	3.15	0.154	2.94
0.0490	22	3.47	0.187	3.52
0.0535	24	3.78	0.222	4.14
0.0579	26	4.10	0.261	4.81
0.0624	28	4.41	0.303	5.51
0.0668	30	4.73	0.347	6.26
0.0713	32	5.04	0.395	7.07
0.0758	34	5.36	0.446	7.92
0.0802	36	5.67	0.500	8.82
0.0847	38	5.99	0.577	9.78
0.0891	40	6.30	0.618	10.79
0.0936	42	6.62	0.681	11.8
0.0980	44	6.93	0.747	12.9
0.102	46	7.25	0.817	14.0
0.107	48	7.56	0.889	15.2
0.111	50	7.88	0.965	16.4
0.123	55	8.67	1.17	19.7
0.134	60	9.46	1.39	23.2
0.145	65	10.24	1.63	27.1
0.156	70	11.03	1.89	31.3
0.167	75	11.8	2.17	35.8
0.178	80	12.6	2.47	40.5
0.189	85	13.4	2.79	45.6
0.201	90	14.2	3.13	51.0
0.212	95	15.0	3.48	56.5
0.223	100	15.8	3.86	62.2
0.245	110	17.3	4.67	74.5
0.267	120	18.9	5.56	88.3
0.290	130	20.5	6.52	103
0.312	140	22.1	7.56	119
0.334	150	23.6	8.68	137

Note: No allowance has been made for age, differences in diameter, or any abnormal condition of interior surface. Any factor of safety must be estimated from the local conditions and the requirements of each particular installation.

Figure 14.2: Friction loss for water in feet per 100 ft. of 1½ inch pipe (Reproduced with the permission of The Hydraulic Institute)

2 INCH NOMINAL		STEEL SCHEDULE 40 ID = 2.067 INCHES ε/D = 0.00087		
DISCHARGE		V	$V^2/2g$	h_f
CFS	GPM	ft/sec	feet	feet per 100 feet of pipe
0.00446	2	0.191	0.000568	0.0151
0.00668	3	0.287	0.00128	0.0302
0.00891	4	0.382	0.00227	0.0497
0.0111	5	0.478	0.00355	0.0731
0.0134	6	0.574	0.00511	0.1004
0.0156	7	0.669	0.00696	0.131
0.0178	8	0.765	0.00909	0.166
0.0201	9	0.860	0.0115	0.205
0.0223	10	0.956	0.0142	0.248
0.0267	12	1.15	0.0205	0.343
0.0312	14	1.34	0.0278	0.453
0.0356	16	1.53	0.0364	0.578
0.0401	18	1.72	0.0460	0.717
0.0446	20	1.91	0.0568	0.868
0.0490	22	2.10	0.0688	1.03
0.0535	24	2.29	0.0818	1.20
0.0579	26	2.49	0.0960	1.39
0.0624	28	2.68	0.111	1.60
0.0668	30	2.87	0.128	1.82
0.0780	35	3.35	0.174	2.42
0.0891	40	3.82	0.227	3.10
0.100	45	4.30	0.288	3.85
0.111	50	4.78	0.355	4.67
0.123	55	5.26	0.430	5.59
0.134	60	5.74	0.511	6.59
0.145	65	6.21	0.600	7.69
0.156	70	6.69	0.696	8.86
0.167	75	7.17	0.799	10.1
0.178	80	7.65	0.909	11.4
0.189	85	8.13	1.03	12.8
0.201	90	8.60	1.15	14.2
0.212	95	9.08	1.28	15.8
0.223	100	9.56	1.42	17.4
0.245	110	10.52	1.72	20.9
0.267	120	11.5	2.05	24.7
0.290	130	12.4	2.40	28.8
0.312	140	13.4	2.78	33.2
0.334	150	14.3	3.20	38.0
0.356	160	15.3	3.64	43.0
0.379	170	16.3	4.11	48.4
0.401	180	17.2	4.60	54.1
0.423	190	18.2	5.13	60.1
0.446	200	19.1	5.68	66.3
0.490	220	21.0	6.88	80.0
0.535	240	22.9	8.18	95.0

Note: No allowance has been made for age, differences in diameter, or any abnormal condition of interior surface. Any factor of safety must be estimated from the local conditions and the requirements of each particular installation.

Figure 14.3: Friction loss for water in feet per 100 ft. of 2 inch pipe (Reproduced with the permission of The Hydraulic Institute)

3 INCH NOMINAL		STEEL SCHEDULE 40 ID = 3.068 INCHES ε/D = 0.000587		
DISCHARGE		V	$V^2/2g$	h_f
CFS	GPM	ft/sec	feet	feet per 100 feet of pipe
0.0111	5	0.217	0.000732	0.0112
0.0223	10	0.434	0.00293	0.0372
0.0334	15	0.651	0.00659	0.0762
0.0446	20	0.868	0.0117	0.126
0.0557	25	1.085	0.0183	0.189
0.0668	30	1.30	0.0263	0.262
0.0780	35	1.52	0.0359	0.347
0.0891	40	1.74	0.0468	0.443
0.100	45	1.95	0.0593	0.547
0.111	50	2.17	0.0732	0.662
0.123	55	2.39	0.0885	0.789
0.134	60	2.60	0.105	0.924
0.145	65	2.82	0.124	1.07
0.156	70	3.04	0.143	1.22
0.167	75	3.25	0.165	1.39
0.178	80	3.47	0.187	1.57
0.189	85	3.69	0.211	1.76
0.201	90	3.91	0.237	1.96
0.212	95	4.12	0.264	2.17
0.223	100	4.34	0.2927	2.39
0.245	110	4.77	0.354	2.86
0.267	120	5.21	0.421	3.37
0.290	130	5.64	0.495	3.92
0.312	140	6.08	0.574	4.51
0.334	150	6.51	0.659	5.14
0.356	160	6.94	0.749	5.81
0.379	170	7.38	0.846	6.53
0.401	180	7.81	0.948	7.28
0.423	190	8.25	1.06	8.07
0.446	200	8.68	1.17	8.90
0.490	220	9.55	1.42	10.7
0.535	240	10.4	1.69	12.6
0.579	260	11.3	1.98	14.7
0.624	280	12.2	2.29	16.9
0.668	300	13.0	2.63	19.2
0.713	320	13.9	3.00	22.0
0.758	340	14.8	3.38	24.8
0.802	360	15.6	3.79	27.7
0.847	380	16.5	4.23	30.7
0.891	400	17.4	4.68	33.9
0.936	420	18.2	5.16	37.3
0.980	440	19.1	5.67	40.9
1.025	460	20.0	6.19	44.6
1.069	480	20.8	6.74	48.5
1.114	500	21.7	7.32	52.5

Figure 14.4: Friction loss for water in feet per 100 ft. of 3 inch pipe (Reproduced with the permission of The Hydraulic Institute)

4 INCH NOMINAL		STEEL SCHEDULE 40 ID = 4.026 INCHES ϵ/D = 0.000447		
DISCHARGE		V	$V^2/2g$	h_f
CFS	GPM	ft/sec	feet	feet per 100 feet of pipe
0.0111	5	0.126	0.000247	0.00310
0.0223	10	0.252	0.000987	0.01017
0.0446	20	0.504	0.00395	0.0344
0.0668	30	0.756	0.00888	0.0702
0.0981	40	1.01	0.0158	0.118
0.111	50	1.26	0.0247	0.176
0.134	60	1.51	0.0355	0.245
0.156	70	1.76	0.0484	0.325
0.178	80	2.02	0.0632	0.415
0.201	90	2.27	0.0800	0.515
0.223	100	2.52	0.0987	0.624
0.245	110	2.77	0.119	0.744
0.267	120	3.02	0.142	0.877
0.290	130	3.28	0.167	1.017
0.312	140	3.53	0.193	1.165
0.334	150	3.78	0.222	1.32
0.356	160	4.03	0.253	1.49
0.379	170	4.28	0.285	1.67
0.401	180	4.54	0.320	1.86
0.423	190	4.79	0.356	2.06
0.446	200	5.04	0.395	2.27
0.490	220	5.54	0.478	2.72
0.535	240	6.05	0.569	3.21
0.579	260	6.55	0.667	3.74
0.624	280	7.06	0.774	4.30
0.668	300	7.56	0.888	4.89
0.713	320	8.06	1.01	5.51
0.758	340	8.57	1.14	6.19
0.802	360	9.07	1.28	6.92
0.847	380	9.58	1.43	7.68
0.891	400	10.1	1.58	8.47
0.936	420	10.6	1.74	9.30
0.980	440	11.1	1.91	10.2
1.025	460	11.6	2.09	11.1
1.069	480	12.1	2.27	12.0
1.114	500	12.6	2.47	13.0
1.225	550	13.9	2.99	15.7
1.337	600	15.1	3.55	18.6
1.448	650	16.4	4.17	21.7
1.560	700	17.6	4.84	25.0
1.671	750	18.9	5.55	28.6
1.782	800	20.2	6.32	32.4
1.894	850	21.4	7.13	36.5
2.005	900	22.7	8.00	40.8
2.117	950	23.9	8.91	45.3

Figure 14.5: Friction loss for water in feet per 100 ft. of 4 inch pipe (Reproduced with the permission of The Hydraulic Institute)

6 INCH NOMINAL		STEEL SCHEDULE 40 ID = 6.065 INCHES ε/D = 0.000293		
DISCHARGE				
CFS	GPM	V ft/sec	$V^2/2g$ feet	h_f feet per 100 feet of pipe
0.0223	10	0.111	0.000192	0.00146
0.0446	20	0.222	0.000767	0.00487
0.0668	30	0.333	0.00172	0.00988
0.0891	40	0.444	0.00307	0.0164
0.111	50	0.555	0.00479	0.0244
0.134	60	0.666	0.00690	0.0337
0.156	70	0.777	0.00939	0.0445
0.178	80	0.888	0.0123	0.0564
0.201	90	0.999	0.0155	0.0698
0.223	100	1.11	0.0192	0.0843
0.267	120	1.33	0.0276	0.118
0.312	140	1.55	0.0376	0.155
0.356	160	1.78	0.0491	0.198
0.401	180	2.00	0.0621	0.246
0.446	200	2.22	0.0767	0.299
0.490	220	2.44	0.0927	0.357
0.535	240	2.66	0.110	0.419
0.579	260	2.89	0.130	0.487
0.624	280	3.11	0.150	0.560
0.668	300	3.33	0.172	0.637
0.713	320	3.55	0.196	0.719
0.758	340	3.78	0.222	0.806
0.802	360	4.00	0.240	0.898
0.847	380	4.22	0.277	0.993
0.891	400	4.44	0.307	1.09
0.936	420	4.66	0.338	1.20
0.980	440	4.89	0.371	1.31
1.025	460	5.11	0.405	1.42
1.07	480	5.33	0.442	1.54
1.11	500	5.55	0.479	1.66
1.23	550	6.11	0.580	1.99
1.34	600	6.66	0.690	2.34
1.45	650	7.22	0.810	2.73
1.56	700	7.77	0.939	3.13
1.67	750	8.33	1.08	3.57
1.78	800	8.88	1.23	4.03
1.89	850	9.44	1.38	4.53
2.01	900	9.99	1.55	5.05
2.12	950	10.5	1.73	5.60
2.23	1 000	11.1	1.92	6.17
2.45	1 100	12.2	2.32	7.41
2.67	1 200	13.3	2.76	8.76
2.90	1 300	14.4	3.24	10.2
3.12	1 400	15.5	3.76	11.8
3.34	1 500	16.7	4.31	13.5
3.56	1 600	17.8	4.91	15.4
3.79	1 700	18.9	5.54	17.3
4.01	1 800	20.0	6.21	19.4
4.23	1 900	21.1	6.92	21.6
4.46	2 000	22.2	7.67	23.8
4.68	2 100	23.3	8.45	26.2
4.90	2 200	24.4	9.27	28.8
5.12	2 300	25.5	10.1	31.4
5.35	2 400	26.6	11.0	34.2
5.57	2 500	27.8	12.0	37.0

Figure 14.6: Friction loss for water in feet per 100 ft. of 6 inch pipe (Reproduced with the permission of The Hydraulic Institute)

8 INCH NOMINAL		STEEL SCHEDULE 40 ID = 7.981 INCHES ε/D = 0.000226		
DISCHARGE		V	$V^2/2g$	h_f
CFS	GPM	ft/sec	feet	feet per 100 feet of pipe
0.0223	10	0.0641	0.0000639	0.000401
0.0446	20	0.128	0.000256	0.001320
0.0668	30	0.192	0.000575	0.00266
0.0891	40	0.257	0.00102	0.00442
0.111	50	0.321	0.00160	0.00652
0.134	60	0.385	0.00230	0.00904
0.156	70	0.449	0.00313	0.01190
0.178	80	0.513	0.00409	0.0151
0.201	90	0.577	0.00518	0.0186
0.223	100	0.641	0.00639	0.0224
0.267	120	0.770	0.00920	0.0311
0.312	140	0.898	0.0125	0.0410
0.356	160	1.03	0.0164	0.0521
0.401	180	1.15	0.0207	0.0644
0.446	200	1.28	0.0256	0.0780
0.490	220	1.41	0.0309	0.0928
0.535	240	1.54	0.0368	0.1088
0.579	260	1.67	0.0432	0.1260
0.624	280	1.80	0.0501	0.144
0.668	300	1.92	0.0575	0.163
0.713	320	2.05	0.0655	0.184
0.758	340	2.18	0.0739	0.206
0.802	360	2.31	0.0828	0.229
0.847	380	2.44	0.0923	0.253
0.891	400	2.57	0.102	0.279
1.003	450	2.89	0.129	0.348
1.11	500	3.21	0.160	0.424
1.23	550	3.53	0.193	0.507
1.34	600	3.85	0.230	0.597
1.45	650	4.17	0.271	0.694
1.56	700	4.49	0.313	0.797
1.67	750	4.81	0.360	0.907
1.78	800	5.13	0.409	1.02
1.89	850	5.45	0.462	1.147
2.01	900	5.77	0.518	1.27
2.12	950	6.09	0.577	1.41
2.23	1 000	6.41	0.639	1.56
2.45	1 100	7.05	0.773	1.87
2.67	1 200	7.70	0.920	2.20
2.90	1 300	8.34	1.08	2.56
3.12	1 400	8.98	1.25	2.95
3.34	1 500	9.62	1.44	3.37
3.56	1 600	10.3	1.64	3.82
3.79	1 700	10.9	1.85	4.29
4.01	1 800	11.5	2.07	4.79
4.23	1 900	12.2	2.31	5.31
4.46	2 000	12.8	2.56	5.86
4.90	2 200	14.1	3.09	7.02
5.35	2 400	15.4	3.68	8.31
5.79	2 600	16.7	4.32	9.70
6.24	2 800	18.0	5.01	11.20
6.68	3 000	19.2	5.75	12.8
7.13	3 200	20.5	6.55	14.5
7.58	3 400	21.8	7.39	16.4
8.02	3 600	23.1	8.28	18.4

Figure 14.7: Friction loss for water in feet per 100 ft. of 8 inch pipe (Reproduced with the permission of The Hydraulic Institute)

10 INCH NOMINAL		STEEL SCHEDULE 40 ID = 10.020 INCHES ε/D = 0.000180		
DISCHARGE		V	$V^2/2g$	h_f
CFS	GPM	ft/sec	feet	feet per 100 feet of pipe
0.0223	10	0.0407	0.0000257	0.000138
0.0446	20	0.0814	0.000103	0.000451
0.0891	40	0.163	0.000412	0.00149
0.134	60	0.244	0.000926	0.00304
0.178	80	0.325	0.00165	0.00505
0.223	100	0.407	0.00257	0.00747
0.267	120	0.488	0.00370	0.0103
0.312	140	0.570	0.00504	0.0136
0.356	160	0.651	0.00659	0.0174
0.401	180	0.732	0.00834	0.0215
0.446	200	0.814	0.0103	0.0260
0.490	220	0.895	0.0125	0.0309
0.535	240	0.976	0.0148	0.0362
0.579	260	1.06	0.0174	0.0417
0.624	280	1.14	0.0202	0.0478
0.668	300	1.22	0.0232	0.0542
0.780	350	1.42	0.0315	0.0719
0.891	400	1.63	0.0412	0.0917
1.003	450	1.83	0.0521	0.114
1.11	500	2.03	0.0643	0.138
1.23	550	2.24	0.0778	0.164
1.34	600	2.44	0.0926	0.192
1.45	650	2.64	0.109	0.224
1.56	700	2.85	0.126	0.256
1.67	750	3.05	0.145	0.291
1.78	800	3.25	0.165	0.328
1.89	850	3.46	0.186	0.368
2.01	900	3.66	0.208	0.410
2.12	950	3.87	0.232	0.455
2.23	1 000	4.07	0.257	0.500
2.45	1 100	4.48	0.311	0.600
2.67	1 200	4.88	0.370	0.703
2.90	1 300	5.29	0.435	0.818
3.12	1 400	5.70	0.504	0.940
3.34	1 500	6.10	0.579	1.07
3.56	1 600	6.51	0.659	1.21
3.79	1 700	6.92	0.743	1.36
4.01	1 800	7.32	0.834	1.52
4.23	1 900	7.73	0.929	1.68
4.46	2 000	8.14	1.03	1.86
4.90	2 200	8.95	1.25	2.23
5.35	2 400	9.76	1.48	2.64
5.79	2 600	10.6	1.74	3.08
6.24	2 800	11.4	2.02	3.56
6.68	3 000	12.2	2.32	4.06
7.13	3 200	13.0	2.63	4.59
7.58	3 400	13.8	2.97	5.16
8.02	3 600	14.6	3.33	5.76
8.47	3 800	15.5	3.71	6.40
8.91	4 000	16.3	4.12	7.07
10.03	4 500	18.3	5.21	8.88
11.1	5 000	20.3	6.43	10.9
12.3	5 500	22.4	7.78	13.2
13.4	6 000	24.4	9.26	15.6
14.5	6 500	26.4	10.9	18.3

Figure 14.8: Friction loss for water in feet per 100 ft. of 10 inch pipe (Reproduced with the permission of The Hydraulic Institute)

12 INCH NOMINAL		STEEL SCHEDULE 40 ID = 11.938 INCHES ε/D = 0.000151		
DISCHARGE		V	$V^2/2g$	h_f
CFS	GPM	ft/sec	feet	feet per 100 feet of pipe
0.223	100	0.287	0.00128	0.00325
0.267	120	0.344	0.00184	0.00448
0.312	140	0.401	0.00250	0.00590
0.356	160	0.459	0.00327	0.00747
0.401	180	0.516	0.00414	0.00920
0.446	200	0.573	0.00511	0.0111
0.490	220	0.631	0.00618	0.0132
0.535	240	0.688	0.00735	0.0155
0.579	260	0.745	0.00863	0.0180
0.624	280	0.802	0.0100	0.0206
0.668	300	0.860	0.0115	0.0233
0.780	350	1.00	0.0156	0.0306
0.891	400	1.15	0.0204	0.0391
1.00	450	1.29	0.0259	0.0485
1.11	500	1.43	0.0319	0.0587
1.23	550	1.58	0.0386	0.0698
1.34	600	1.72	0.0460	0.0820
1.45	650	1.86	0.0539	0.0950
1.56	700	2.01	0.0626	0.109
1.67	750	2.15	0.0718	0.124
1.78	800	2.29	0.0817	0.140
1.89	850	2.44	0.0922	0.156
2.01	900	2.58	0.103	0.173
2.12	950	2.72	0.115	0.191
2.23	1 000	2.87	0.128	0.210
2.45	1 100	3.15	0.154	0.251
2.67	1 200	3.44	0.184	0.296
2.90	1 300	3.73	0.216	0.344
3.12	1 400	4.01	0.250	0.395
3.34	1 500	4.30	0.287	0.450
3.56	1 600	4.59	0.327	0.509
3.79	1 700	4.87	0.369	0.572
4.01	1 800	5.16	0.414	0.636
4.23	1 900	5.45	0.461	0.704
4.46	2 000	5.73	0.511	0.776
4.90	2 200	6.31	0.618	0.930
5.35	2 400	6.88	0.735	1.093
5.79	2 600	7.45	0.863	1.28
6.24	2 800	8.03	1.00	1.47
6.68	3 000	8.60	1.15	1.68
7.13	3 200	9.17	1.31	1.90
7.58	3 400	9.75	1.48	2.13
8.02	3 600	10.3	1.65	2.37
8.47	3 800	10.9	1.84	2.63
8.91	4 000	11.5	2.04	2.92
10.03	4 500	12.9	2.59	3.65
11.1	5 000	14.3	3.19	4.47
12.3	5 500	15.8	3.86	5.38
13.4	6 000	17.2	4.60	6.39
14.5	6 500	18.6	5.39	7.47
15.6	7 000	20.1	6.26	8.63
16.7	7 500	21.5	7.18	9.88
17.8	8 000	22.9	8.17	11.20
18.9	8 500	24.4	9.22	12.6
20.1	9 000	25.8	10.3	14.1

Figure 14.9: Friction loss for water in feet per 100 ft. of 12 inch pipe (Reproduced with the permission of The Hydraulic Institute)

14.2 Typical resistance coefficients for valves and fittings

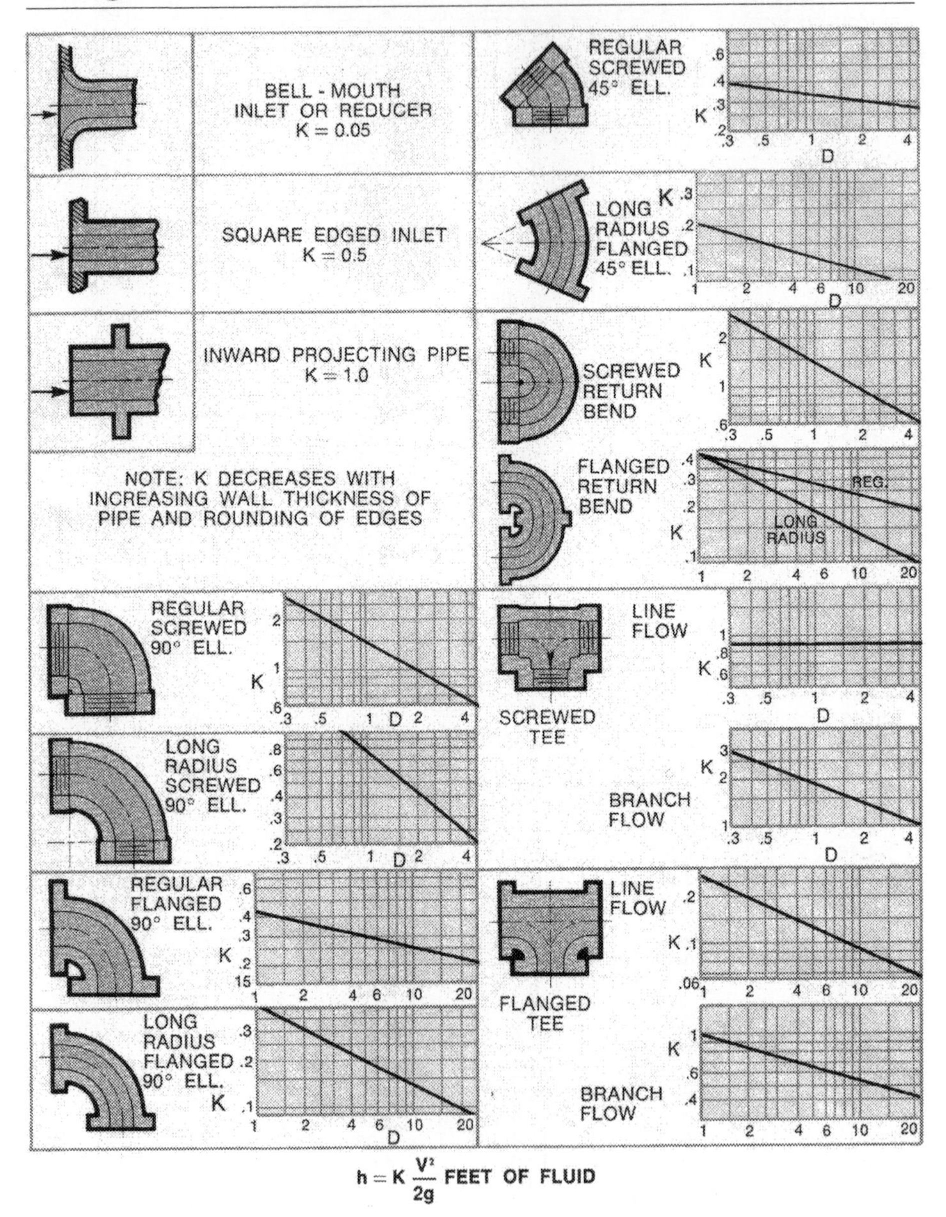

$$h = K \frac{V^2}{2g} \text{ FEET OF FLUID}$$

Figure 14.10: Typical resistance coefficients for valves and fittings (Reproduced with the permission of The Hydraulic Institute)

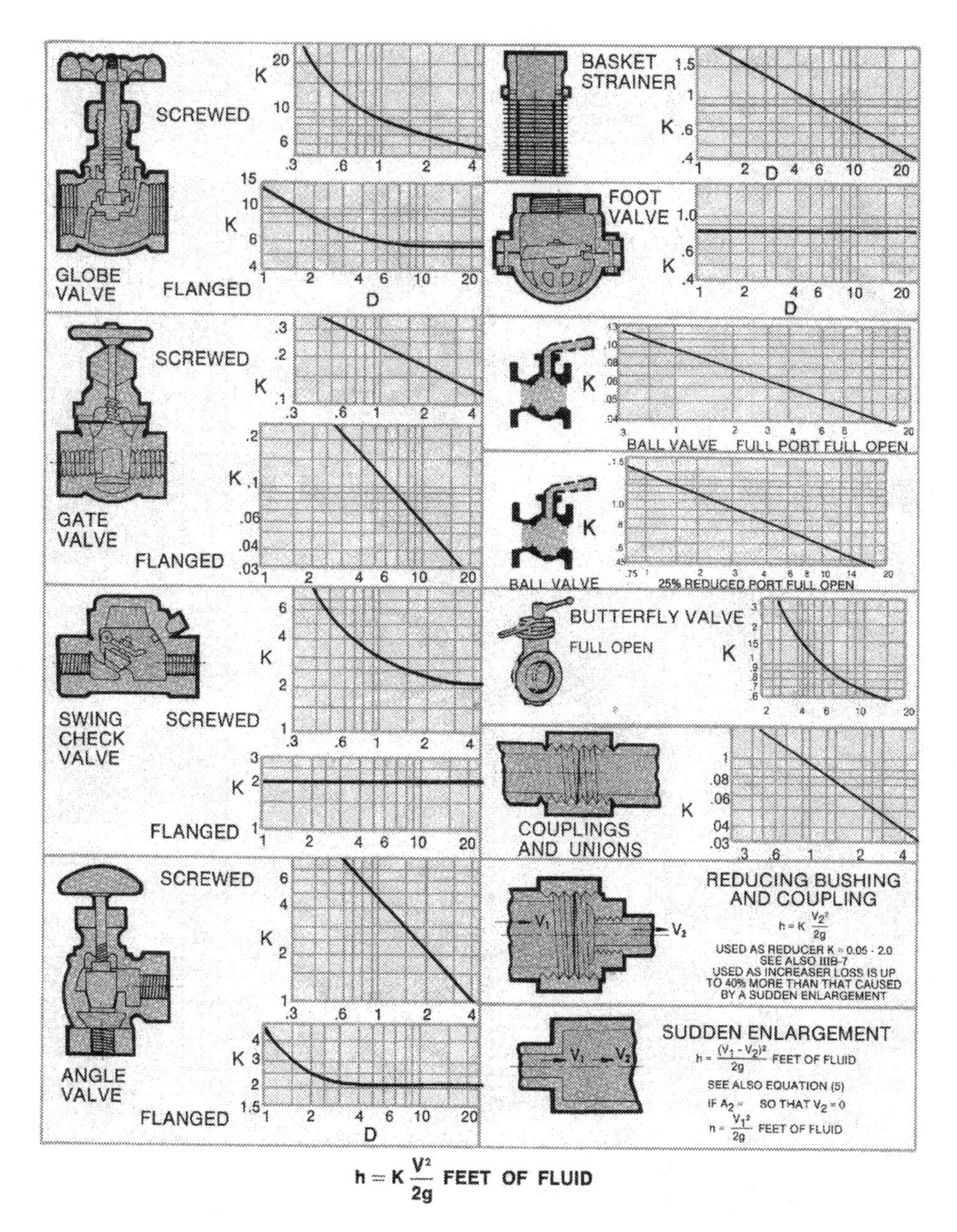

$$h = K \frac{V^2}{2g} \text{ FEET OF FLUID}$$

Figure 14.11: Typical resistance coefficients for valves and fittings (Reproduced with the permission of The Hydraulic Institute)

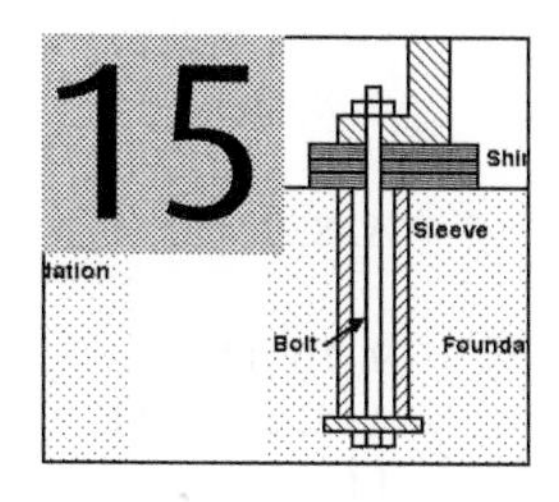

Materials of construction

15.1 Materials of construction for pumping various liquids

The material selections listed in the last column are those that are commonly used for pumping the liquids identified. Further details of these material selections are identified in Chapter 15.5.1.

Liquid	Conditions of liquid	Chemical symbol	Specific gravity	Material selection
Acetaldehyde		C_2H_4O	0.78	C
Acetate solvents				A, B, C, 8, 9, 10, 11
Acetone		C_3H_6O	0.79	B, C
Acetic anhydride		$C_2H_6O_3$	1.08	8, 9, 10, 11, 12
Acide, acetic	Conc. cold	$C_2H_4O_2$	1.05	8, 9, 10, 11, 12
Acid, acetic	Dil. cold			A, 8, 9, 10, 11, 12
Acid acetic	Conc. boiling			9, 10, 11, 12
Acid acetic	Dil. boiling			9, 10, 1, 12
Acid, arsenic, ortho		H_2AsO_4, 1/2 H_2O	2.0 – 2.5	8, 9, 10, 11, 12
Acid, benzoic		$C_7H_6O_2$	1.27	8, 9, 10, 11
Acid, boric	Aqueous sol.	H_3BO_3		A, 8, 9, 10, 11, 12
Acid butyric	Conc.	$C_4H_8O_2$	0.96	8, 9, 10, 11
Acid, carbolic	Conc. (M.P.106°F)	C_6H_6O	1.07	C, 8, 9, 10, 11
Acid, carbolic	(see phenol)			B, 8, 9, 10, 11
Acid, carbolic	Aqueous soln.	$CO_2 + H_2O$		A
Acid, chromic	Aqueous soln.	$Cr_2O_3 + H_2O$		A, 8, 9, 10, 11, 12

Liquid	Conditions of liquid	Chemical symbol	Specific gravity	Material selection
Acid, citric	Aqueous soln.	$C_6H_8O_7$ + H_2O		A, 8, 9, 10, 11, 12
Acids, fatt (oleic palmiticx, stearic, etc.)				A, 8, 9, 10, 11
Acid, formic		CH_2O_2	1.22	9, 10, 11
Acid, fruit				A, 8, 9, 10, 11, 14
Acid, hydrochloric	Coml. conc.	HCL	1.19 (38%)	11, 12
Acid, hydrochloric	Dil. cold			10, 11, 12, 14, 15
Acid, hydrochloric	Dil. hot			11, 12
Acid, hydrocyanic		HCN	0.7	C, 8, 9, 10, 11
Acid, hydrofluoric	Anhydrous, with hydrocarbon	HF + H_xC_x		3, 14
Acid, hydrofluoric	Aqueous soln.	HF		A, 14
Acid hydrofluosilcic		H_2SiF_6	1.3	A, 14
Acid, lactic		C3H6O3	1.25	A, 8, 9, 10, 11, 12
Acid, mine water				A, 8, 9, 10, 11
Acid, mixed	Sulfuric + nitric			C, 3, 8, 9, 10, 11, 12
Acid, muriatic	(see acid, hydrochloric)			
Acid, naphthenic				C, 5, 8, 9, 10, 11
Acid, nitric	Conc. boiling	HNO3	1.5	6, 7, 10, 12,
Acid, nitric	Dilute			5, 6, 7, 8, 9, 10, 12
Acid, oxalic	Cold	$C_2H_2O_4.2H_2O$	1.65	8, 9, 10, 11, 12
Acid, oxalic	Hot	$C_2H_2O_4.2H_2O$		10, 11, 12
Acid, ortho-phosphoric		H3PO4	1.87	9, 10, 11
Acid, picric		C6H3N3O7	1.76	8, 9, 10, 11, 12
Acid, pyrogallic		C6H6O3	1.45	8, 9, 10, 11
Acid, pyroligneous				A 8, 9, 10, 11
Acid, sulfuric	>77% cold	H_2SO_4	1.69 – 1.94	C, 10, 11, 12
Acid, sulfuric	65/93% >175°F			11, 12
Acid, sulfuric	65/93% <175°F			10, 11, 12
Acid, sulfuric	10 – 65%			10, 11, 12
Acid, sulfuric	10%			A, 10, 11, 12, 14
Acid, sulfuric(Oleum)	Fuming	H_2SO_4 + SO3	1.92–1.94	3, 10, 11
Acid sulfurous		H_2SO 3		A 8, 9, 10, 11

Liquid	Conditions of liquid	Chemical symbol	Specific gravity	Material selection
Acid, tannic		$C_{14}H_{10}O_9$		A, 8, 9, 10, 11, 14
Acid, tartaric	Aqueous soln.	$C_4H_6O_6 \cdot H_2O$		A, 8, 9, 10, 11, 14
Alcohols				A, B
Alum	(see aluminum sulphate and potash alum)			
Aluminium sulphate	Aqueous soln.	$AL_2(SO_4)_3$		10, 11, 12, 14,
Ammonia, aqua		NH_4OH		C
Ammonium bicarbonate	Aqueous soln.	NH_4HCO_3		C
Ammonium chloride	Aqueous soln.	NH_4Cl		9, 10, 11, 12, 14
Ammonium nitrate	Aqueous soln.	NH_4NO_3		C, 8, 9, 10, 11, 14,
Ammonium phosphate dibasic	Aqueous soln.	$(NH_4)_2HPO_4$		C, 8, 9, 10, 11, 14
Ammonium sulfate	Aqueous soln.	$(NH_4)_2SO_4$		C, 8, 9, 10, 11
Aniline		C_6H_7N	1.02	B, C
Aniline hydrochloride	Aqueous soln.	$C_6H_5NH_2HCl$		11, 12
Asphalt	Hot		0.98 – 1.4	C, 5
Barium chloride	Aqueous soln.	$BaCl_2$		C, 8, 9, 10, 11
Barium nitrate	Aqueous soln.	$Ba(NO_3)_2$		C, 8, 9, 10, 11
Beer				A, 8
Beer wort				A, 8
Beet juice				a, 8
Beet pulp				A, B, 8, 9, 10, 11
Benzene		C_6H_6	0.88	
Benzine	(see petroleum ether)			
Benzol	(see benzene)			B, C,
Bichloride of mercury	(see mercuric chloride)			
Black liquor	(see liquor, pulp mill)			
Bleach solutions	(see type)			
Blood				A, B
Brine				
(calcium chloride)	pH > 8	$CaCl_2$		C
(calcium chloride)	pH < 8			A, 10, 11, 13, 14

Liquid	Conditions of liquid	Chemical symbol	Specific gravity	Material selection
Brine *contd.*				
(calcium and magnesium chloride)	Aqueous soln.			A, 10, 11, 13, 14
(calcium and sodium chloride)	Aqueous soln.			A, 10, 11, 13, 14
(sodium chloride)	Under 3% salt, cold	NaCl		A, C, 13
(sodium chloride)	Over 3% salt, cold		1.02 1.2	A, B, 9, 10, 11, 13, 14
(sodium chloride)	Over 3% salt, hot			9, 10, 11, 12, 14
Brine, seawater			1.03	A, B, C
Butane		C4H10	0.5 at 0oC	B, C, 3
Calcium bisulfite	Paper mill	$Ca(HSO_3)_2$	1.06	9, 10, 11
Calcium chlorate	Aqueous soln.	$CA(CIO_3)_2$ $2H_20$		10, 11, 12
Calcium hypochlorite		$Ca(OCl)_2$		C, 10, 11, 12
Cane juice				A, B, 13
Carbon bisulfide		CS_2	1.26	C
Carbonate of soda	(see soda ash)			
Carbon tetrachloride	Anhydrous	CCl_4	1.5	B, C
Carbon tetrachloride	Plus water			A, 8
Catsup				A, 8, 9, 10, 11
Caustic potash	(see potassium hydroxide)			
Caustic soda	(see sodium hydroxide)			
Cellulose acetate				9, 10, 11
Chlorate of lime	(see calcium chlorate)			
Chloride of lime	(see calcium hypochlorite)			
Chlorine water	(depending on conc.)			
Chlorobenzene		C_6H_5Cl	1.1	A, B, 8
Chloroform		CHCl3	1.5	A, 8, 9, 10, 11, 14
Chrome alum	Aqueous soln.	$CrK(SO_4)_2 \cdot 12H_2O$		10, 11, 12
Condensate	(see water, distilled)			

Liquid	Conditions of liquid	Chemical symbol	Specific gravity	Material selection
Copperas, green	(see ferrous sulphate)			
Copper ammonium acetate	Aqueous soln			C, 8, 9, 10, 11
Copper chloride (cupric)	Aqueous soln.	$CuCl_2$		11, 12
Copper nitrate		Cu(NO3)2		8, 9, 10, 11
Copper sulfate blue vitriol	Aqueous soln.	$CuSO_4$		8, 9, 10, 11, 12
Creosote	(see oil, creosote)			
Cresol, meta		C_7H_8O	1.03	C, 5
Cyanide	(see sodium cyanide and potassium cyanide)			
Cyanogen	In water	$(CN)_2$ gas		C
Diphenyl		$C_6H_5 . C_6H_5$	0.99	C, 3
Enamel				C
Ethanol	(see alcohols)			
Ethylene chloride (di-chloride)	Cold	$C_2H_4Cl_2$	1.28	A, 8, 9, 10, 11, 14
Ferric chloride	Aqueous soln.	$FeCl_3$		11, 12
Ferric sulphate	Aqueous soln.	$Fe_2(SO_4)_3$		8, 9, 10, 11, 12
Ferrous chloride	Cod, Aqueous	$FeCl_2$		11, 12
Ferrous sulphate (gree copperass	Aqueous soln	$FeSO_4$		9, 10, 11, 12, 14
Formaldehyde		CH2)	1.08	A, 8, 9, 10, 11
Fruit juices				A, 8, 9, 10, 11, 14
Furfural		$C_5H_4O_2$	1.16	A, C, 8, 9, 10, 11
Gasoline			0.68–0.75	B, C.
Gaubes salt	(see sodium sulfate)			
Glucose				A, B
Glue	Hot			B, C
Glue sizing				A
Glycerol (glycerin)		$C_3H_8O_3$	1.26	A, B, C
Green liquor	(see liquor, pulp mill)			
Heptane		C_7H_{16}	0.69	B, C
Hydrogen peroxide	Aqueous soln	H_2O_2		8, 9, 10, 11

Liquid	Conditions of liquid	Chemical symbol	Specific gravity	Material selection
Hydrogen sulfide	Aqueous soln	H_2S		81 91 10, 11
Hydrosulfite of soda	(see sodium hydrosulfite)			
Hyposulfite of soda	(see sodium thiosulfate)			
Kaoline slip	Suspension in water			C, 3
Kaoline slip	Suspension in acid			10, 11, 12
Kerosene	(see oil kerosene)			
Lard	Hot			B, C
Lead acetate (sugar of lead)	Aqueous soln.	$Pb(C_2H_3O_2)_2.3H_2O$		9, 10, 11, 14
Lead	Molten			C, 3
Lime water (milk of lime)		Ca(OH)2		C
Liquor-pulp mill; black				C, 3, 9, 10, 11, 12, 14
Liquor-pulp mill; green				C, 3, 9, 10, 11, 12, 14
Liquor-pulp mill; white				C, 3, 9, 10, 11, 12, 14
Liquor-pulp mill; pink				C, 3, 9, 10, 11, 12, 14
Liquor-pulp mill; sulfite				9, 10, 11
Lithium chloride	Aqueous soln.	LiCl		C
Lye, caustic	(See potassium and sodium hydroxide)			
Magnesium chloride	Aqueous soln.	$MgCl_2$		10, 11, 12
Magnesium sulfate (epson salts)	Aqueous soln.	$MgSO_4$		C, 8, 9, 10, 11
Manganese chloride	Aqueous soln.	$MnCl_2.4H_2O$		A, 8, 9, 10, 11, 12
Manganeous sulfate	Aqueous soln.	$MnSO_4.4H_2O$		A, C, 8, 9, 10, 11
Mash				A, B, 8
Mercuric chloride	Very dilute aqueous sol	$HgCl_2$		9, 10, 11, 12
Mercuric chloride	Coml conc. aqueous sol	$HgCl_2$		11, 12
Mercuric sulfate	In sulfuric acid	$HgSO_4 + H_2SO_4$		10, 11, 12
Mercurous sulfate	In sulfuric acid	$HgSO_4 + H_2SO_4$		10, 11, 12

Liquid	Conditions of liquid	Chemical symbol	Specific gravity	Material selection
Methyl chloride		CH_3Cl	0.52	C
Methylene chloride		CH_2Cl_2	1.34	C, 8
Milk			1.03–1.04	8
Milk of lime	(See lime water)			
Mine water	(See acid, mine water)			
Miscella	(20% soybean oil & solvent)		0.75	C
Molasses				A, B
Mustard				A, 8, 9, 10, 11, 12
Naphtha			0.78–0.88	B, C
Naphtha, crude			0.92–0.95	B, C
Nicotine sulfate		$(C_{10}H_{14}N_2)_2 H_2SO_4$		10, 11, 12, 14
Nitre	(See potassium nitrate)			
Nitre cake	(See sodium bisulphate)			
Nitro ethane		$C_2H_5NO_2$	1.04	B, C
Nitro methane		CH_3NO_2	1.14	B, C
Oil, coal tar				B, C, 8, 9, 10, 11
Oil, coconut			0.91	A, B, C, 8, 9, 10, 11, 14
Oil, creosote			1.04–1.1	B, C
Oil, crude	Cold			B, C,
Oil, crude	Hot			3
Oil, essential				A, B, C
Oil, fuel				B, C,
Oil, kerosene				B, C
Oil, linseed			0.94	A, B, C, 8, 9, 10, 11, 14
Oil, lubricating				B, C
Oil, mineral				B, C
Oil, olive			0.9	B, C
Oil, palm			0.9	A, B, C, 8, 9, 10, 11, 14
Oil, quenching			0.91	B, C
Oil, rapeseed			0.92	A, 8, 9, 10, 11, 14
Oil, soya bean				A, B, C, 8, 9, 10, 11, 14

Liquid	Conditions of liquid	Chemical symbol	Specific gravity	Material selection
Oil, turpentine			0.87	B, C
Paraffin	Hot			B, C
Perhydrol	(See hydrogen peroxide)			
Peroxide of Hydrogen	(See hydrogen peroxide)			
Petroleum ether				B, C
Phenol		C_6H_6O	1.07	
Pink liquor	(See liquor, pulp mill)			
Photographic developers				8, 9, 10, 11
Plating solutions	(Varied and complicated, consult pump manufacturers)			
Potash	Plant liquor			A, 8, 9, 10, 11, 13, 14
Potash alum	Aqueous soln.	$Al_2(SO_4)_3K_2SO_4.24H_2O$		A, 9, 10, 11, 12, 13, 14
Potassium bichromate	Aqueous soln.	$K_2Cr_2O_7$		C
Potassium carbonate	Aqueous soln.	K_2CO_3		C
Potassium chlorate	Aqueous soln.	$KClO_3$		8, 9, 10, 11, 12
Potassium chloride	Aqueous soln.	KCl		A, 8, 9, 10, 11, 14
Potassium cyanide	Aqueous soln.	KCN		C
Potassium hydroxide	Aqueous soln.	KOH		C, 5, 8, 9, 10, 11, 13, 14
Potassium nitrate	Aqueous soln.	KNO_3		C, 5, 8, 9, 10, 11
Potassium sulfate	Aqueous soln.	K_2SO_4		A, 8, 9, 10, 11
Propane		C_3H_8	0.59 @48°	F B, C, 3
Pyridine		C_5H_5N	0.98	C
Pyridine sulfate				10, 12
Rhidolene				B
Rosin (colophony)	Paper mill			C
Sal ammoniac	(See ammonium chloride)			
Salt lake	Aqueous soln.	Na_2SO_4 + impurities		A, 8, 9, 10, 11, 12
Salt water	(See brines)			
Sea water	(See brines)			

Liquid	Conditions of liquid	Chemical symbol	Specific gravity	Material selection
Sewage				A, B, C
Shellac				A
Silver nitrate	Aqueous soln.	$AgNO_3$		8, 9, 10, 11, 12
Slop, brewery				A, B, C
Slop, distillers				A, 8, 9, 10, 11
Soap liquor				C
Soda ash	Cold	Na_2CO_3		C
Soda ash	Hot			8, 9, 10, 1, 13, 14
Sodium bicarbonate	Aqueous soln.	Na_2HCO_3		C, 8, 9, 10, 11, 13
Sodium bisulfate	Aqueous soln.	$NaHSO_4$		10, 11, 12
Sodium carbonate	(See soda ash)			
Sodium chlorate	Aqueous soln.	$NaClO_3$		8, 9, 10, 11, 12
Sodium chloride	(See brines)			
Sodium cyanide	Aqueous soln.	$NaCN$		C
Sodium hydroxide	Aqueous soln.	$NaOH$		C, 5, 8, 9, 10, 11, 13, 14
Sodium Hydrosulfite	Aqueous soln.	$Na_2S_2O_4$, $2H_2O$		8, 9, 10, 11
Sodium hypochlorite		$NaOCl$		10, 11, 12
Sodium hyposulfite	(See sodium thiosulfate)			
Sodium meta silicate				C
Sodium nitrate	Aqueous soln.	$NaNO_3$		C, 5, 8, 9, 10, 11
Sodium phosphate: monobasic	Aqueous soln.	$NaH_2PO_4.H_2O$		A., 8, 9, 10, 11
Sodium phosphate: dibasic	Aqueous soln.	$Na_2HPO_4.7H_2O$		A, C, 8, 9, 10, 11
Sodium phosphate: tribasic	Aqueous soln.	$Na_2HPO_4.12H_2O$		C
Sodium phosphate: meta	Aqueous soln.	$Na_4P_4O_{12}$		A, 8, 9, 10, 11
Sodium phosphate: hexameta	Aqueous soln.	$(NaPO_3)_6$		8, 9, 10, 11
Sodium plumbite	Aqueous soln.			C
Sodium sulfate	Aqueous soln.	Na_2SO_4		A, 8, 9, 10, 11
Sodium sulfide	Aqueous soln.	Na_2S		C, 8, 9, 10, 11
Sodium sulfite	Aqueous soln.	Na_2SO_3		A, 8, 9, 10, 11
Sodium thiosulfate	Aqueous soln.	$Na_2S_2O_3.5H_2O$		8, 9, 10, 11
Stannic chloride	Aqueous soln.	$SnCl_4$		11, 12

Liquid	Conditions of liquid	Chemical symbol	Specific gravity	Material selection
Starch	(C6H10O5)x			A, B
Strontium nitrate	Aqueous soln.	$Sr(NO_3)_2$		C, 8
Sugar	Aqueous soln.			A, 8, 9, 10, 11, 13
Sulfite liquor	(See liquor, pulp mill)			
Sulfur	In water	S		A, C, 8, 9, 10, 11
Sulfur	Molten	S		C
Sulfur chloride	Cold	S_2Cl_2		C
Syrup	(See sugar)			
Tallow	Hot		0.99	C
Tanning liquors				A, B, 9, 10, 11, 12, 14
Tar	Hot			C, 3
Tar and ammonia	In water			C
Tetrachloride of tin	(See tannic chloride)			
Tetraethyl lead		$Pb(C_2H_5)_4$	1.66	B, C
Toluene (toluol)		C_7H_8	0.87	B, C
Trichloroethylene		C_2HCl_3	1.47	A, B, C, 8
Urine				A, 8, 9, 10, 11
Varnish				A, B, C, 8, 14
Vegetable juices				A, 8, 9, 10, 11, 14
Vinegar				A, 8, 9, 10, 11, 12
Vitriol, blue	(See copper sulfate)			
Vitriol, green	(See ferrous sulfate)			
Vitriol, oil of	(See acid, sulfuric)			
Vitriol, white	(See zinc sulfate)			
Water, boiler feed	Not evaporated pH>8.5		1.0	C
Water, boiler feed	High makeup pH<8.5			B
Water, boiler feed	Low makeup Evaporated		1.0	4, 5, 8, 14
Water, distilled	High purity		1.0	A, B
Water, distilled	Condensate			A, B

Liquid	Conditions of liquid	Chemical symbol	Specific gravity	Material selection
Water, fresh			1.0	B
Water, mine	(See acid, mine water)			
Water, salt and sea	(See brines)			
Whiskey				A, B
White liquor	(See liquor, pulp mill)			
White water	Paper mill			A, B, C
Wine				A, B
Wood pulp (stock)				A, B, C
Wood vinegar	(See acid pyroligneous)			
Wort	(Se beer wort)			
Xylol (Xylene)		C_8H_{10}	0.87	B, C, 8, 9, 10, 11
Yeast				A, B
Zinc chloride	Aqueous soln.	$ZnCl_2$		9, 10, 11, 12
Zinc sulfate	Aqueous soln.	$ZnSO_4$		A, 9, 10, 11

Reproduced with permission of the Hydraulic Institute

15.2 Material selection

Materials	ASTM Number	Remarks
A		All bronze construction
B		Bronze fitted construction
C		All iron construction
3	A216–WCB	Carbon steel
4	A217–C5	5% chromium steel
5	A743–CA15	12% chromium steel
6	A743–CB30	20% chromium steel
7	A743–CC50	28% chromium steel
8	A743–CF–8	19–9 austenitic steel
9	A743–CF–8M	19–10 molybdenum austenitic steel
10	A743–CN–7M	20–29 chromium nickel austenitic steel with copper and molybdenum
11		A series of nickel-based alloys
12	A518	Corrosion-resistant high silicon cast iron
13	A436	Austenitic cast iron – 2 types
13(a)	A439	Ductile austenitic cast iron
14		Nickel-copper alloy
15		Nickel

Reproduced with permission of the Hydraulic Institute

15.3 Material classes for centrifugal pumps in general refinery services

CAUTION: This table is intended as a general guide. It should not be used without a knowledgeable review of the specific services involved

Service	On-Plot Process Reference Plant	Off-Plot Transfer & Loading	Temperature Range (°F)	Pressure Range (psig)	Material Class	See Note
Fresh Water, condensate, cooling tower water	X	X	<212	All	I–1 or I–2	
Boiling Water and process water	X	X	<250	All	I–1 or I–2	5
	X	X	250–350	All	S–5	5
	X	X	>350	All	C–6	5
Boiler Feed Water						
Axially split	X	X	>200	All	C–6	
Double casing (barrel)	X	X	>200	All	S–6	
Boiler Circulator	X	X	>200	All	C–6	
Foul water, reflux, drum water water draw, and hydrocarbons containing these	X	X	<350	All	S–3 or S–66	
waters, including reflux streams	X	X	>350	All	C–6	
Propane, butane, liquified petroleum gas and ammonia	X	X	<450	All	S–1	
Diesel Oil; gasoline; naphtha; kerosene; gas oils; light, medium and	X	X	<450	All	S–1	
heavy lube oils; fuled oil; residuum; crude oil; asphalt; synthetic crude	X		450–700	All	S–6	6,7
bottoms	X		>700	All	C–6	6

Service	On-Plot Process Reference Plant	Off-Plot Transfer & Loading	Temperature Range (°F)	Pressure Range (psig)	Material Class	See Note
Noncorrosive hydrocarbons, e.g. catalytic reformate, ixomataxate, desulfurized oils.	X	X	45–700	All	S–4	7
Xylene, toluene, acetone, benzene furfural, MEK, cumene	X	X	<450	All	S–1	
Sodium carbonate, doctor solution	X	X	<350	All	I-1	
Caustic (sodium hydroxide, concentration of ≤ 20%	X	X	<210	All	S–1	8
			≥210	All		9
Seawater	X	X	<200	All	–	10
MEA, DEA, TEA – stock solutions	X	X	<250	All	S–1	
DEA, TEA – lean solutions	X	X	<250	All	S–1	8
MEA-lean solution (CO_2 only)	X	X	175–300	All	S–9	8
MEA-lean solution (CO_2 and H_2S)	X	X	175–300	All		8,11
MEA, DEA, TEA – rich solutions	X	X	<175	All	S–1	8
Sulfuric Acid concentrations						
85%	X	X	<100	All	S–1	6
85% – 15%	X	X	<100	All	A–8	6
15% – 1%	X	X	<100	All	A–8	6
1%	X	X	<450	All	A–8	6
Hydrofluoric acid concentration of >96%	X	X	<100	All	S–9	6

General notes

1. The materials for pump parts for each material class are given in Appendix H.
2. Separate materials recommendations should be obtained for services not clearly identified by the service descriptions listed in this table.
3. Cast iron casings, where recommended for chemical services, are for nonhazardous locations only. Steel casings (S-1 or I-1) should be used for pumps in services located near process plants or in any location where released vapor from a failure could create a hazardous situation or where pumps could be subjected to hydraulic shock, for example, in loading service.
4. Mechanical seal materials: for streams containing chlorides, all springs and other metal parts should be Alloy 20 or better. Buna-N and Neoprene should not be used in any service containing aromatics. Viton should not be used in services containing aromatics above 200°F

Reference notes

5. Oxygen content and buffering of water should be considered in the selection of material.
6. The corrosiveness of foul waters, hydrocarbons over 450°F, acids and acid sludges may vary widely. A materials recommendation should be obtained for each service. The material class indicated above will be satisfactory for many of these services but must be verified.
7. If product corrosivity is low, Class S–4 materials may be used for services at 451°F–700°F. A separate materials recommendation should be obtained in each instance.
8. For temperatures greater than or equal to 160°F, all welds should be stress relieved.
9. Alloy 20 or Monel pump material and double mechanical seals should be used with a pressurized seal oil system.
10. For seawater service, the purchaser and the vendor should agree on the construction materials that best suit the intended use.
11. Class A–7 materials should be used, except for carbon steel casings.

15.4 Materials for pump parts

Table H-1 - Materials for Pump Parts

Part	Full[b] Compliance Material?	Material Class and Material Abbreviations[a]												
		I-1	I-2	S-1	S-3	S-4	S-5	S-6	S-8	S-9	C-6	A-7	A-8	D-1
		CI / CI	CI / BRZ	STL / CI	STL / NI-RESIST	STL / STL	STL / STL 12% CHR	STL / 12% CHR	STL / 316 AUS	STL / MONEL	12% CHR / 12% CHR	AUS / AUS (1&2)	316AUS / 316 AUS(1&2)	DUPLEX / DUPLEX
Pressure Casing	Yes	Cast iron	Cast iron	Carbon steel	Carbon steel	Carbon steel	Carbon steel	Carbon steel	Carbon steel	Carbon steel	12% CHR	AUS	316 AUS	Duplex
Inner case parts (bowls, diffusers, diaphragms)	No	Cast iron	Bronze	Cast iron	Ni-resist	Cast iron	Carbon steel	12% CHR	316 AUS	Monel	12% CHR	AUS	316 AUS	Duplex
Impeller	Yes	Cast iron	Bronze	Cast iron	Ni-resist	Carbon steel	Carbon steel	12% CHR	316 AUS	Monel	12% CHR	AUS	316 AUS	Duplex
Case wear rings	No	Cast iron	Bronze	Cast iron	Ni-resist	Cast iron	12% CHR	12% CHR	Hard Faced 316 AUS (3)	Monel	12% CHR hardened	Hard Faced AUS (3)	Hard Faced 316 AUS (3)	Duplex (3)
Impeller wear rings	No	Cast iron	Bronze	Cast iron	Ni-resist	Cast iron	12% CHR Hardened	12% CHR Hardened	Hard faced 316 AUS (3)	Monel	12% CHR hardened	Hard Faced AUS (3)	Hard Faced 316 AUS (3)	Duplex (3)
Shaft (2)	Yes	Carbon steel	Carbon steel	Carbon steel	Carbon steel	Carbon steel	AISI 4140	AISI 4140 (4)	316 AUS	K-Monel	12% CHR	AUS	316 AUS	Duplex
Shaft sleeves, packed pumps	No	12% CHR hardened	Hard bronze	12% CHR hardened	12% CHR hardened or hard faced	12% CHR hardened or hard faced	12% CHR hardened or hard faced	12% CHR hardened or hard faced	Hard Face 316 AUS (3)	K-Monel, hardened	12% CHR hardened or hard faced	Hard Faced AUS (3)	Hard Faced 316 AUS (3)	Duplex (3)
Shaft sleeves, mechanical seals	No	AUS or 12% CHR	AUS or 12% CHR	AUS or 12% CHR	AUS or 12%CHR	AUS or 12% CHR	AUS or 12% CHR	AUS or 12% CHR	AUS or 12% CHR	K-Monel, hardened	AUS or 12% CHR	AUS	316 AUS	Duplex
Throat bushings	No	Cast iron	Bronze	Cast iron	Ni-resist	Cast iron	12% CHR	12% CHR	316 AUS	Monel	12% CHR hardened	AUS	316 AUS	Duplex
Interstage sleeves	No	Cast iron	Bronze	Cast iron	Ni-resist	Cast iron	12% CHR hardened	12% CHR hardened	Hard Faced 316 AUS (3)	K-Monel, hardened	12% CHR hardened	Hard Faced AUS (3)	Hard Faced 316 AUS (3)	Duplex (3)
Interstage bushings	No	Cast iron	Bronze	Cast iron	Ni-resist	Cast iron	12% CHR hardened	12% CHR hardened	Hard Faced 316 AUS (3)	K-Monel, hardened	12% CHR hardened	Hard Faced AUS (3)	Hard Faced 316 AUS (3)	Duplex (3)
Seal gland	Yes	316 AUS (5)	316 AUS (5)	316 AUS (5)	316 AUS (5)	316 AUS (5)	316 AUS (5)	316 AUS (5)	316 AUS (5)	Monel	316 AUS (5)	316 AUS (5)	316 AUS (5)	Duplex (5)
Case and gland studs	Yes	Carbon steel	Carbon steel	AISI 4140 steel	AISI 4140 steel	AISI 4140 steel	AISI 4140 steel	AISI 4140 steel	AISI 4140 steel	K-Monel, hardened (8)	AISI 4140 steel	AISI 4140 steel	AISI 4140 steel	Duplex (8)
Case gasket	No	AUS, spiral wound (6)	AUS, spiral wound (6)	AUS, spiral wound (6)	AUS, spiral wound (6)	AUS, spiral wound (6)	AUS, spiral wound (6)	AUS, spiral wound (6)	316 AUS, spiral wound (6)	Monel, spiral wound, PTFE filled (6)	AUS, spiral wound (6)	AUS, spiral wound (6)	316 AUS spiral wound (6)	Duplex SS spiral wound (6)
Discharge head / suction can	Yes	Carbon steel	Carbon steel	Carbon steel	Carbon steel	Carbon steel	Carbon steel	Carbon steel	Carbon steel	Carbon steel	AUS	AUS	316 AUS	Duplex
Column / bowl shaft bushings	No	Nitrile (7)	Bronze	Filled carbon	Nitrile (7)	Filled carbon	Filled carbon	Filled carbon	Filled carbon	Filled carbon	Filled carbon	Filled carbon	Filled carbon	Filled carbon
Wetted fasteners (bolts)	Yes	Carbon steel	Carbon steel	Carbon steel	Carbon steel	Carbon steel	316 AUS	316 AUS	316 AUS	K-Monel	316 AUS	316 AUS	316 AUS	Duplex

a The abbreviation above the diagonal line indicates the case material; the abbreviation below the diagonal line indicates trim material.
Abbreviations are as follows: BRZ = bronze, STL - steel, 12% CHR = 12% chrome, AUS = austenitic stainless steel, CI = cast iron, 316 AUS = Type 316 austenitic stainless steel

b See 2.11.1.1

Figure 15.1: Materials for pump parts (Reproduced with permission of the American Petroleum Institute)

15.5 Material specifications for pump parts

Table H.2 — Materials specifications for pump parts

Material Class	Applications	USA ASTM	USA UNS[a]	International ISO	Europe EN[b]	Europe Grade	Japan Material No	Japan JIS
Cast Iron	Pressure Castings	A 278 Class 30	F 12401	185/ Gr. 250	EN 1561	EN-GJL-250	JL 1040	G 5501, FC 300
	General Castings	A 48 Class 25/30/40	F 11 701/ F 12 101	185/ Gr. 300	EN 1561	EN-GJL-250 EN-GJL-300	JL 1040 JL 1050	G 5501, FC 250/300
Carbon Steel	Pressure Castings	A 216 Gr WCB	J 03 002	4991 C23-45 AH	EN 10213-2	GP 240 GH	1.0619	G 5151, Cl SCPH 2
	Wrought / Forgings	A 266 Class 2	K 03506	683-18-C25	EN 10222-2	P 280 GH	1.0426	G 3202, Cl SFVC 2A
	Bar Stock: Pressure	A 696 Gr B40	G 10 200	683-18-C 25	EN 10273	P 295 GH	1.0481	G 4051, Cl S25C
	Bar Stock: General	A 576 Gr 1045	G 10 450	683-18-C45e	EN 10083-2	C 45	1.0503	G 4051, Cl S45C
	Bolts and Studs (General)	A 193 Gr B7	G 41 400	2604-2-F31	EN 10269	42 Cr Mo 4	1.7225	G 4107, Cl SNB7
	Nuts(General)	A 194 Gr 2H	K 04 002	683-1-C35e	EN 10269	C 35 E	1.1181	G 4051, Cl S45C
	Plate	A 516 Gr 65/70	K 02 403/ K 02 700		EN 10028-2	P 295 GH	1.0481	G 3106, Gr SM400B
	Pipe	A 106 GrB	K 03 006		EN 10208-1	L 245 GA	1.0459	G 3456, Gr. STPT 370/410
	Fittings	A 105	K 03 504					G 4051, Cl S25C G 3202, Cl SFVC 2A, SFVC2B
AISI 4140 Steel	Bar Stock	A 434 Class BB A 434 Class BC	G 41 400[c]		EN 10083-1	42 Cr Mo 4	1.7225	G 4105, C1 SCM 440
	Bolts and Studs	A 193 Gr B7	G 41 400		EN 10269	42 Cr Mo 4	1.7225	G 4107, C1 SNB16
	Nuts	A 194 Gr 2H	K 04 002	2604-2-F31	EN 10269	C 35 E	1.1181	G 4051, C1 S45C

Figure 15.2: Material specifications for pump parts (Reproduced with permission of the American Petroleum Institute)

Table H.2— Materials specifications for pump parts (continued)

Material Class	Applications	USA		International	Europe		Japan	
		ASTM	UNS [a]	ISO	EN [b]	Grade	Material No	JIS
12% Chrome Steel	Pressure Castings	A 217 Gr CA 15	J 91 150		EN 10213-2	GX 8 Cr Ni 12	1.4107	G 5121, C1 SCS 1
		A 487 Gr CA6NM	J 91 540		EN 10213-2	GX 4 Cr Ni 13-4	1.4317	G 5121, C1 SCS 6
	General Castings	A 743 Gr CA 15	J 91 150		EN 10283	GX 12 Cr 12	1.4011	
		A 743 Gr CA6NM	J 91 540		EN 10283	GX 4 Cr Ni 13-4	1.4317	
	Wrought / Forgings: Pressure	A 182 Gr F6a Cl 1 A 182 Gr F 6 NM	S 41 000 S 41 500	683-13-3	EN 10250-4 EN 10222-5	X12 Cr 13 X 3 Cr NiMo 13-4-1	1.4006 1.4313	G 3214, C1 SUS F6 B G 3214, C1 SUS F6NM
	Wrought / Forgings: General	A 473 Type 410	S 41 000	683-13-2	EN 10088-3	X 12 Cr 13	1.4006	G 3214, C1 SUS F6 NM
	Bar Stock: Pressure	A 479 Type 410	S 41 000	683-13-3	EN 10272	X12 Cr 13	1.4006	G 4303 or 410
	Bar Stock: General	A 276 Type 410	S 41 400	683-13-3	EN 10088-3	X 12 Cr 13	1.4006	G 4303, Grl SUS 403 or 420
	Bar Stock: Forgings [c]	A 276 Type 420 A 473 Type 416 A 582 Type 416	S 42 000 S 41 600 S 41 600	683-13-4	EN 10088-3	X 20 Cr 13 X 20 Cr S 13 X 20 Cr S 13	1.4021 1.4005 1.4005	G 4303, Grl SUS 403 or 420
	Bolts and Studs [d]	A 193 Gr B6	S 41 000	3506-1, C4-70	EN 10269	X22CrMoV 12-1	1.4923	G 4303, Gr SUS 403 or 420
	Nuts [d]	A 194 Gr 6	S 41 000	3506-2, C4-70	EN 10269	X22CrMoV 12-1	1.4923	G 4303, Gr SUS 403 or 420
	Plate	A 240 Type 410	S 41 000	683-13-3	EN 10088-2	X 12 Cr 13	1.4006	G 4304, 410

Figure 15.2: Material specifications for pump parts *continued*

Table H.2— Materials specifications for pump parts (continued)

Material Class	Applications	USA ASTM	USA UNS[a]	International ISO	Europe EN[b]	Europe Grade	Japan Material No	Japan JIS
Austenitic Stainless Steel	Pressure Casting s	A 351 Gr CF3	J 92 500	683-13-10	EN 10213-4	GX2 Cr Ni 19-11	1.4309	G 5121, C1 SCSI 3A
		A 351 Gr CF3M	J 92 800	683-13-19	EN 10213-4	GX2 Cr Ni Mo 19-11-2	1.4409	G 5121, C1 SCS I 4A 02M
	General Castings	A 743 Gr CF3	J 92 500		EN 10283	GX2 Cr Ni 19-11	1.4309	
		A 743 Gr CF3M	J 92 800		EN 10283	GX2 Cr Ni Mo 19-11-2	1.4409	
	Wrought / Forgings	A 182 Gr F 304L	S 30 403	683-13-10	EN 10222-5 EN 10250-4	X2 Cr Ni 19-11	1.4306	G 3214, C1 SUS F 304 L
		A 182 Gr F 316L	S 31 603	683-13-19	EN 10222-5 EN 10250-4	X2 Cr Ni Mo 17-12-2	1.4404	G 3214, C1 SUS F 316 L
	Bar Stock [f]	A 479 Type 304L A 479 Type 316L	S 30 403 S 31 603	683-13-10 683-13-19	EN 10088-3 EN 10088-3	X2 Cr Ni 19-11 X2 Cr Ni Mo 17-12-2	1.4306 1.4404	G 4303, SUS F 304L G 4303, SUS F 316L
		A 479 Type XM19 [e]	S 20910					
	Plate	A 240 Gr 304L / 316L	S 30 403 S 31 603	683-13-10 683-13-19	EN 10028-7 EN 10028-7	X2 Cr Ni 19-11 X2 Cr Ni Mo 17-12-2	1.4306 1.4404	G 4304, Gr 304L/ 316L
	Pipe	A 312 Type 304L 316L	S 30 403 S 31 603	683-13-10 683-13-19				G 3459, Gr. SUS 304LTP/316LTP
	Fittings	A 182 Gr F304L Gr 316L	S 30 403 S 31 603	683-13-10 683-13-19	EN 10222-5	X2 Cr Ni 19-11 X2 Cr Ni Mo 17-12-2	1.4306 1.4404	G 3214, Dr SUS 304L 316L
	Bolts and Studs	A 193 Gr B 8 M	S 31 600	683-1-21	EN 10250-4	X6 Cr Ni Mo Ti 17-12-2	1.4571	G 4303, Grl SUS 316
	Nuts	A 194 Gr B 8 M	S 31 600	683-1-21	EN 10250-4	X6 Cr Ni Mo Ti 17-12-2	1.4571	G 4303, Grl SUS 316

Figure 15.2: Material specifications for pump parts *continued*

Table H.2— Materials specifications for pump parts (continued)

Material Class	Applications	USA		International	Europe		Japan	
		ASTM	UNS[a]	ISO	EN[b]	Grade	Material No	JIS
Duplex Stainless Steel	Pressure Castings	A 351 Gr CD4 MCu A 890 Gr 1 B	J93370 J 93372		EN 10213-4	GX2 CrNiMoCuN 25-6-3-3	1.4517	
		A 890 Gr 3A	J93371					G 5121, Gr. SCS 11
		A 890 Gr 4A	J92205		EN 10213-4	GX2 CrNiMoCuN 25-6-3-3	1.4517	
	Wrought / Forgings	A 182 Gr F 51 A 479	S 31803 S 32550		EN 10250-4 EN 10222-5 EN 10088-3	X2 Cr Ni Mo N 22-5-3 X2 Cr Ni Mo Cu N 25-6-3	1.4462 1.4507	G 4319, Cl SUS 329
	Bar Stock	A 276-S31803	S 31 803		EN 10088-3	X2 Cr Ni Mo N 22-5-3	1.4462	G 4303, Grl SUS 329 Gr SUS 329
	Plate	A 240-S31803	S 31 803		EN 10028-7	X2 Cr Ni Mo N 22-5-3	1.4462	G 4303, Gr SUS 329
	Pipe	A 790-S31803	S 31 803					G 3459, Gr. SUS 329J3LTP
	Fittings	A 182 G r F 51	S 31803		EN 10250-4 EN 10222-5	X2 Cr Ni Mo N 22-5-3	1.4462	
	Bolts and Studs	A 276-S31803	S 31 803		EN 10088-3	X2 Cr Ni Mo N 22-5-3	1.4462	G 4303, Gr SUS 329
	Nuts	A 276-S31803	S 31 803		EN 10088-3	X2 Cr Ni Mo N 22-5-3	1.4462	G 4303, Gr SUS 329

Figure 15.2: Material specifications for pump parts *continued*

Table H.2— Materials specifications for pump parts (continued)

Material Class	Applications	USA		International	Europe		Japan	
		ASTM	UNS[a]	ISO	EN[b]	Grade	Material No	JIS
Super Duplex Stainless Steel[g]	Pressure Castings	A 351 Gr CD3MWCuN	J93380					
		A 890 Gr 5A	J93404		EN 10213-4	GX2 Cr Ni Mo N 26-7-4	1.4469	
		A 890 Gr 6A	J93380					
	Wrought / Forgings	A 182 Gr 55	S 32 760		EN 10250-4 EN 10088-3	X2 Cr Ni Mo Cu WN 25-7-4	1.4501	
	Bar Stock	A 276-S32760 A 479-S32760	S 32 760		EN 10088-3	X2 Cr Ni Mo Cu WN 25-7-4	1.4501	
	Plate	A 240-S32760	S 32 760		EN 10028-7	X2 Cr Ni Mo Cu WN 25-7-4	1.4501	
	Pipe	A 790-S32760	S 32 760					G 3459, Gr. SUS 329 J4LTP
	Fittings	A 182 Gr F55	S 32 760		EN 10250-4 EN 10088-3	X2 Cr Ni Mo Cu WN 25-7-4	1.4501	
	Bolts and Studs	A 276-S32760	S 32 760		EN 10088-3	X2 Cr Ni Mo Cu WN 25-7-4	1.4501	
	Nuts	A 276-S32760	S 32 760		EN 10088-3	X2 Cr Ni Mo Cu WN 25-7-4	1.4501	

[a] UNS (unified numbering system) designation for chemistry only.

[b] Where EN standards do not yet exist, there are available European national standards, e.g. AFNOR, BS, DIN, etc.

[c] Do not use for shafts in the hardened condition (over 302 HB).

[d] Special, normally use AISI 4140.

[e] Nitronic 50 or equivalent.

[f] For shafts, standard grades of 304 and 316 may be substituted in place of low carbon (L) grades

[g] Super duplex stainless steel classified with Pitting Resistance Equivalent (PRE) number greater than or equal to 40

PRE = $\%Cr_{free}$ + (3,3 x % Molybdenum) + (2 x % Copper) + (2 x %Tungsten) + (16 x % Nitrogen)

= [(%Chromium - (14,5 x %Carbon)] + (3,3 x %Molybdenum) + (2 x %Copper) + (2 x %Tungsten) + (16 x %Nitrogen)

Figure 15.2: Material specifications for pump parts *continued*

16 Conversion tables and formulae

16.1 Conversion tables

To convert multiply first column by the second column to obtain the result in the third column.

16.1.1 Length

Feet	0.3048	Meters
Feet	30.48	Centimeters
Feet	12	Inches
Inches	2.54	Centimeters
Inches	25.4	Millimeters
Kilometers	1000	Meters
Kilometers	0.6214	Miles
Kilometers	1094	Yards
Meters	100	Centimeters
Meters	3.281	Feet
Meters	39.37	Inches
Meters	0.001	Kilometers
Meters	1.094	Yards
Miles	280	Feet
Miles	1.609	Kilometers
Miles	1760	Yards
Millimeters	0.1	Centimeters

Millimeters	0.03937	Inches
Yards	91.44	Centimeters
Yards	3	Feet
Yards	36	Inches
Yards	0.9144	Meters

16.1.2 Area

Acres	43,560	Square Feet
Acres	4047	Square Meters
Acres	4840	Square Yards
Acres	0.4047	Hectares
Square Centimeters	0.1550	Square Inches
Square Centimeters	100	Square Millimeters
Square Feet	0.0929	Square Meters
Square Feet	929	Square Centimeters
Square Feet	144	Square Inches
Square Inches	6.452	Square centimeters
Square Inches	645.2	Square Millimeters
Square Kilometers	247.1	Acres
Square Kilometers	1,000,000	Square Meters
Square Meters	10.76	Square Feet
Square Meters	1.196	Square Yards
Square Miles	2.59	Square Kilometers
Square Yards	9	Square Feet
Square Yards	0.8361	Square Meters

16.1.3 Volume

Barrels of Oil	42	US Gallons of Oil
Barrels of Beer	31	US Gallons of Beer
Barrels of Whisky	45	US Gallons of Whisky
Cubic Feet	1728	Cubic Inches
Cubic Feet	0.02832	Cubic Meters
Cubic Feet	0.03704	Cubic Yards

Cubic Feet	7.48052	US Gallons
Cubic Feet	28.32	Liters
Cubic Feet	59.84	Pints
Cubic Feet	29.92	Quarts
Cubic Inches	0.03463	Pints
Cubic Inches	0.01732	Quarts
Cubic Inches	16387	Cubic Millimeters
Cubic Meters	1,000,000	Cubic Centimeters
Cubic Meters	35.31	Cubic Feet
Cubic Meters	1.308	Cubic Yards
Cubic Meters	264.2	US Gallons
Cubic Meters	999.97	Liters
Cubic Meters	2113	Pints
Cubic Meters	1057	Quarts
Cubic Yards	764,554.86	Cubic Centimeters
Cubic Yards	27	Cubic Feet
Cubic Yards	46,656	Cubic Inches
Cubic Yards	0.7646	Cubic Meters
Cubic Yards	1616	Pints
Cubic Yards	807.9	Quarts
Imperial Gallons	1.20095	US Gallons
Liters	1000	Cubic Centimeters
Liters	0.03531	Cubic Feet
Liters	61.02	Cubic Inches
Liters	0.2642	US Gallons
Liters	2.113	Pints
Liters	1.057	Quarts
Quarts	57.75	Cubic Inches
Quarts	0.9464	Liters
US Gallons	3785	Cubic Centimeters
US Gallons	0.1337	Cubic Feet

US Gallons	231	Cubic Inches
US Gallons	3.785	Liters
US Gallons	8	Pints
US Gallons	4	Quarts
US Gallons	0.83267	Imperial Gallons
US Gallons water	8.345	Pounds of water

16.1.4 Velocity

Centimeters per Second	1.969	Feet per Minute
Centimeters per Second	0.03281	Feet per Second
Centimeters per Second	0.036	Kilometers per Hour
Centimeters per Second	0.6	Meters per Minute
Centimeters per Second	0.02237	Miles per Hour
Feet per Minute	0.5080	Centimeters per Second
Feet per Minute	0.01667	Feet per Second
Feet per Minute	0.01829	Kilometers per Hour
Feet per Minute	0.00508	Meters per Second
Feet per Minute	0.01136	Miles per Hour
Feet per Second	30.48	Meters per Minute
Feet per Second	1.097	Kilometers per Hour
Feet per Second	0.5924	Knots
Feet per Second	0.3048	Meters per Second
Feet per Second	0.6818	Miles per Hour
Feet per Second	0.01136	Miles per Minute
Kilometers per Hour	27.78	Centimeters per Second
Kilometers per Hour	54.68	Feet per Minute
Kilometers per Hour	0.9113	Feet per Second
Kilometers per Hour	0.5399	Knots
Kilometers per Hour	16.67	Meters per Minute
Kilometers per Hour	0.6214	Miles per Hour
Meters per Minute	1.667	Centimeters per Second
Meters per Minute	3.281	Feet per Minute

Meters per Minute	0.05468	Feet per Second
Meters per Minute	0.06	Kilometers per Hour
Meters per Minute	0.03728	Miles per Hour
Meters per Second	196.8	Feet per Minute
Meters per Second	3.281	Feet per Second
Meters per Second	3.6	Kilometers per Hour
Meters per Second	0.06	Kilometers per Minute
Meters per Second	2.287	Miles per Hour
Miles per Hour	44.70	Centimeters per Second
Miles per Hour	88	Feet per Minute
Miles per Hour	1.467	Feet per Second
Miles per Hour	1.609	Kilometers per Hour
Miles per Hour	0.8689	Knots
Miles per Hour	26.82	Meters per Minute
Miles per Minute	2682	Centimeters per Second
Miles per Minute	88	Feet per Second
Miles per Minute	1.609	Kilometers per Minute
Miles per Minute	60	Miles per Hour

16.1.5 Capacity

Barrels per Day of Oil	0.02917	US Gallons per Minute of Oil
Cubic Feet per Minute	472.0	Cubic Centimeters per Second
Cubic Feet per Minute	0.02832	Cubic Meters per Minute
Cubic Feet per Minute	0.1247	US Gallons per Second
Cubic Feet per Minute	0.4719	Liters per Second
Cubic Feet per Minute	62.43	Pounds of Water per Minute
Cubic Feet per Second	0.646317	Millions Gallons per Day
Cubic Feet per Second	448.831	US Gallons per Minute

Cubic Meters per Hour	0.016667	Cubic Meters per Minute
Cubic Meters per Hour	4.4033	US Gallons per Minute
Cubic Meters per Second	15,850	US Gallons per Minute
Tons of Water per Day	83.333	Pounds of Water per Hour
Tons of Water per Day	0.16643	US Gallons per Minute
Tons of Water per Day	1.3349	Cubic Feet per Hour
US Gallons per Minute	0.06308	Liters per Second
US Gallons per Minute	3.785	Liters per Minute
US Gallons per Minute	8.0208	Cubic Feet per Hour

16.1.6 Pressure or Head

Atmospheres	1.01325	Bars
Atmospheres	76.0	Centimeters of Mercury
Atmospheres	29.92	Inches of Mercury
Atmospheres	33.90	Feet of Water
Atmospheres	1.0332	Kilograms per Square Centimeter
Atmospheres	10,332	Kilograms per Square Meter
Atmospheres	101.325	Kilopascals
Atmospheres	14.7	Pounds per Square Inch
Bars	100	Kilopascal
Bars	0.9869	Atmospheres
Bars	1.0197	Kilograms per Square Centimeter
Bars	14.504	Pounds per Square Inch
Feet of Water	0.0295	Atmospheres
Feet of Water	0.8826	Inches of Mercury
Feet of Water	304.8	Kilograms per Square Meter
Feet of Water	0.0304	Kilograms per Square Centimeter

Feet of Water	62.43	Pounds per Square Foot
Feet of Water	0.4335	Pounds per Square Inch
Kilograms per Square Meter	0.2048	Pounds per Square Foot
Kilograms per Square Meter	9.807	Pascal
Kilograms per Square Centimeter	0.9678	Atmospheres
Kilograms per Square Centimeter	0.9807	Bars
Kilograms per Square Centimeter	32.87	Feet of Water
Kilograms per Square Centimeter	98.066	Kilopascals
Kilograms per Square Centimeter	14.223	Pounds per Square Inch
Kilopascals	0.3351	Feet of Water
Kilopascals	0.0102	Kilograms per Square Centimeter
Kilopascals	1,000	Pascals
Kilopascals	0.145	Pounds per Square Inch
Pounds per Square Inch	0.06804	Atmospheres
Pounds per Square Inch	6895	Pascal
Pounds per Square Inch	2.307	Feet of Water
Pounds per Square Inch	2.036	Inches of Mercury
Pounds per Square Inch	703.1	Kilograms per Square Meter
Pounds per Square Inch	0.0703	Kilograms per Square Centimeter
Pounds per Square Inch	6.8948	Kilopascals

16.1.7 Energy, Work, Heat

British Thermal Units	0.2520	Kilogram Calories
British Thermal Units	777.6	Foot Pounds
British Thermal Units	107.5	Kilogram Meters

British Thermal Units	1055	Joules
Calorie	4.1868	Joules
Foot Pounds Force	0.1383	Kilogram Meters
Foot Pounds Force	1.356	Joules
Foot Pounds Force	0.3238	Calories
Horsepower Hours	2546	British Thermal Units
Horsepower Hours	1,980,000	Foot Pounds
Horsepower Hours	641.6	Kilogram Calories
Horsepower Hours	0.7457	Kilowatt Hours
Joules	0.7376	Foot Pounds Force
Kilowatt Hours	3414.4	British Thermal Units
Kilowatt Hours	1.341	Horsepower Hours
Kilowatt Hours	860.4	Kilogram Calories

16.1.8 Power

B.T.U. per Second	1055	Watts
B.T.U. per Minute	12.96	Foot pounds per Second
B.T.U. per Minute	0.02356	Horsepower
B.T.U. per Minute	0.01757	Kilowatts
B.T.U. per Minute	17.57	Watts
Foot Pounds per Minute	0.01667	Foot Pounds per Second
Horsepower	42.44	B.T.U. per Minute
Horsepower	33,000	Foot Pounds per Minute
Horsepower	550	Foot Pounds per Second
Horsepower	10.547	Kilogram Calories per Minute
Horsepower	0.7457	Kilowatts
Kilogram Calorie per Second	3.968	B.T.U. per Second
Kilogram Calorie per Second	3086	Foot Pounds per Second
Kilogram Calorie per Second	5.6145	Horsepower

Kilogram Calorie per Second	4186.7	Watts
Kilogram Calorie per Minute	0.09351	Horsepower
Kilogram Calorie per Minute	69.733	Watts
Kilowatts	56.907	B.T.U. per Minute
Kilowatts	737.6	Foot Pounds per Second
Kilowatts	1.341	Horsepower
Kilowatts	14.34	Kilogram Calories per Minute
Kilowatts	1,000	Watts

16.1.9 Mass

Grams	0.001	Kilograms
Grams	0.03527	Ounces
Kilograms	2.2046	Pounds
Ounces	0.0625	Pounds
Ounces	28.35	Grams
Pounds	16	Ounces
Pounds	0.45359	Kilogram
Tons (short	2000	Pounds
Tons (short)	0.9072	Metric Ton (or Tonne)
Tons (short)	907.2	Kilograms
Tons (long)	2240	Pounds
Tons (long)	1.016	Metric Ton (or Tonne)
Tons (long)	1016	Kilograms

16.2 Useful formulae

16.2.1 Head-pressure relationship

$$\text{Pressure (in p.s.i.)} = \frac{\text{Head (in feet)} \times \text{Specific Gravity}}{2.31}$$

$$\text{Pressure (in Kg/Sq.Cm)} = \frac{\text{Head (in meters)} \times \text{Specific Gravity}}{10.2}$$

16.2.2 Power calculation

$$\text{H.P.} = \frac{\text{USGPM} \times \text{Head (in feet)} \times \text{Specific Gravity}}{\text{Efficiency} \times 3960}$$

$$\text{kW} = \frac{\text{m}^3\text{/hr.} \times \text{Head (in meters)} \times \text{Specific Gravity}}{\text{Efficiency} \times 367.12}$$

16.2.3 The affinity laws

$$\frac{Q_2}{Q_1} = \frac{D_2}{D_1} \quad \text{or} \quad \frac{N_2}{N_1}$$

$$\frac{H_2}{H_1} = \left(\frac{D_2}{D_1}\right)^2 \quad \text{or} \quad \left(\frac{N_2}{N_1}\right)^2$$

$$\frac{HP_2}{HP_1} = \left(\frac{D_2}{D_1}\right)^3 \quad \text{or} \quad \left(\frac{N_2}{N_1}\right)^3$$

16.2.4 Pressure in a static head system

$$P = \frac{Hs \times s.g.}{2.31}$$

where P = Gauge pressure in pounds per square inch

Hs = Static Head of liquid in feet

s.g. = Specific Gravity of the liquid being pumped

16.2.5 Pressure in a flowing system

$$P = (Hs - Hf) \times \frac{s.g.}{2.31}$$

where P = Gauge pressure in pounds per square inch

Hs = Static Head of liquid in feet

Hf = Friction Head in feet

s.g. = Specific Gravity of the liquid being pumped

16.2.6 The Effect of capacity change on pipe friction

$$\left[\frac{Q_2}{Q_1}\right]^2 = \frac{Hf_2}{Hf_1}$$

where Q = Flow Rate

Hf = Friction Loss

16.2.7 The effect of head change on the flow rate

$$\left[\frac{Q_2}{Q_1}\right]^2 = \frac{Hs_2}{Hs_1}$$

where Q = Flow Rate

Hs = Static Head

16.2.8 The effect of changes in pipe diameter on the friction in a system

$$\left[\frac{D_1}{D_2}\right]^5 = \frac{Hf_2}{Hf_1}$$

where D = Pipe Diameter

Hf = Friction Loss

16.2.9 The effect of changes in pipe diameter on the flow rate in a system

$$\left[\frac{D_2}{D_1}\right]^{2.5} = \frac{Q_2}{Q_1}$$

where D = Pipe Diameter

Q = Flow Rate

16.2.10 Total head in a system

$$\text{Total Head (H)} = Hs + Hsd + Hv + Hf$$

where Hs = Static Head

Hsd = System Differential Head

Hv = Increase in Velocity Head across pump

Hf = Friction Losses

16.2.11 Net positive suction head available from a system

$$NPSHA = Hs + Ha - Hvp - Hf$$

where Hs = Static Head over impeller

Ha = Surface Pressure in source

Hvp = Vapor Pressure of liquid

Hf = Friction Losses

16.2.12 Impeller specific speed

$$Ns = \frac{N\sqrt{Q}}{H^{0.75}}$$

where Ns = Specific Speed

N = Pump Rotational Speed

Q = Flow Rate at BEP
(50% of BEP with double suction impeller)

H = Total Head at BEP

16.2.13 Suction specific speed

$$Nss = \frac{RPM \times Q^{0.5}}{NPSHR^{0.75}}$$

where RPM = Pump Rotational Speed

Q = Flow at BEP

NPSHR = NPSH required at BEP

16.2.14 Calculation of grease service life

$$T = K \left[\frac{14.0 \times 106}{n \times \sqrt{d}}\right] - 4d$$

where T = Grease service life (hours)

K = Factor for bearing type (10 for ball brgs.)

n = Pump rotational speed

d = Bearing Bore (mm)

16.2.15 L-10 Bearing life calculation

$$L\text{-}10 = A_{23} \left[\frac{C}{P}\right]^{p} \times \frac{1{,}000{,}000}{60 \times n}$$

where a_{23} = Lubrication effectiveness factor

C = Basic dynamic load rating

P = Equivalent dynamic bearing load

p = Exponent of the life equation (3 for ball brgs.)

p = Pump rotational speed

16.2.16 Slurry specific gravity

$$S_m = \frac{S_L}{1 + Cw\left[\frac{S_L}{S_s} - 1\right]}$$

where S_m = Specific Gravity of the Slurry

S_L = Specific Gravity of the Liquid

S_s = Specific Gravity of the Solids

Cw = Concentration of solids by weight (30% is used as 0.30)

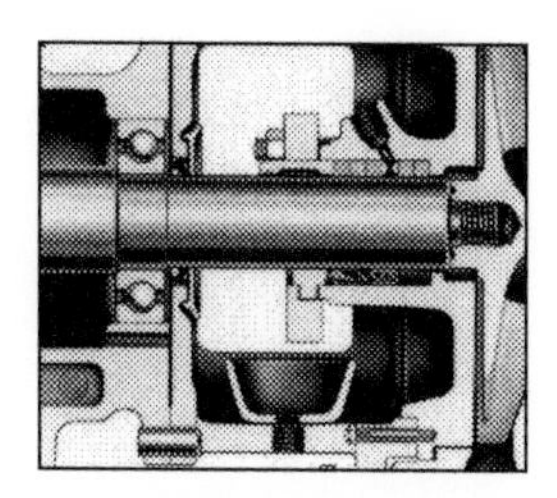

Bibliography

12 Steps to Mechanical Seal Reliability in Centrifugal Pumps, Ross Mackay Associates Ltd., 2002

American Petroleum Institute Standard 610, Ninth Edition, 2003

ANSI/HI 2000 Edition Pump Standards, Hydraulic Institute

Cameron Hydraulic Data, Ingersoll-Dresser Pump Co., 1994

Centrifugal Pump Sourcebook, John Dufour & William E. Nelson, McGraw Hill, Inc., 1993

Pump Engineering Manual, The Duriron Company, Inc., 1980

Pump Handbook, Igor J. Karassik, et al., McGraw-Hill, 1976

Pumping Station Design, Robert L. Sanks, et al., Butterworth-Heinemann, 1998

The Pump Handbook Series, Pumps & Systems, 1993

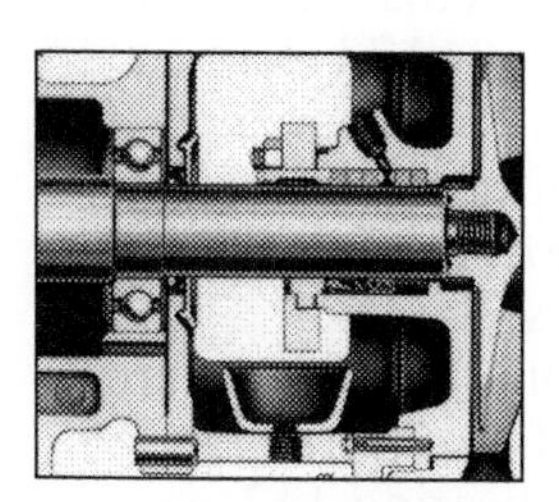

Index

Printed and bound by CPI Group (UK) Ltd, Croydon, CR0 4YY

08/05/2025

01864850-0004